D1123026

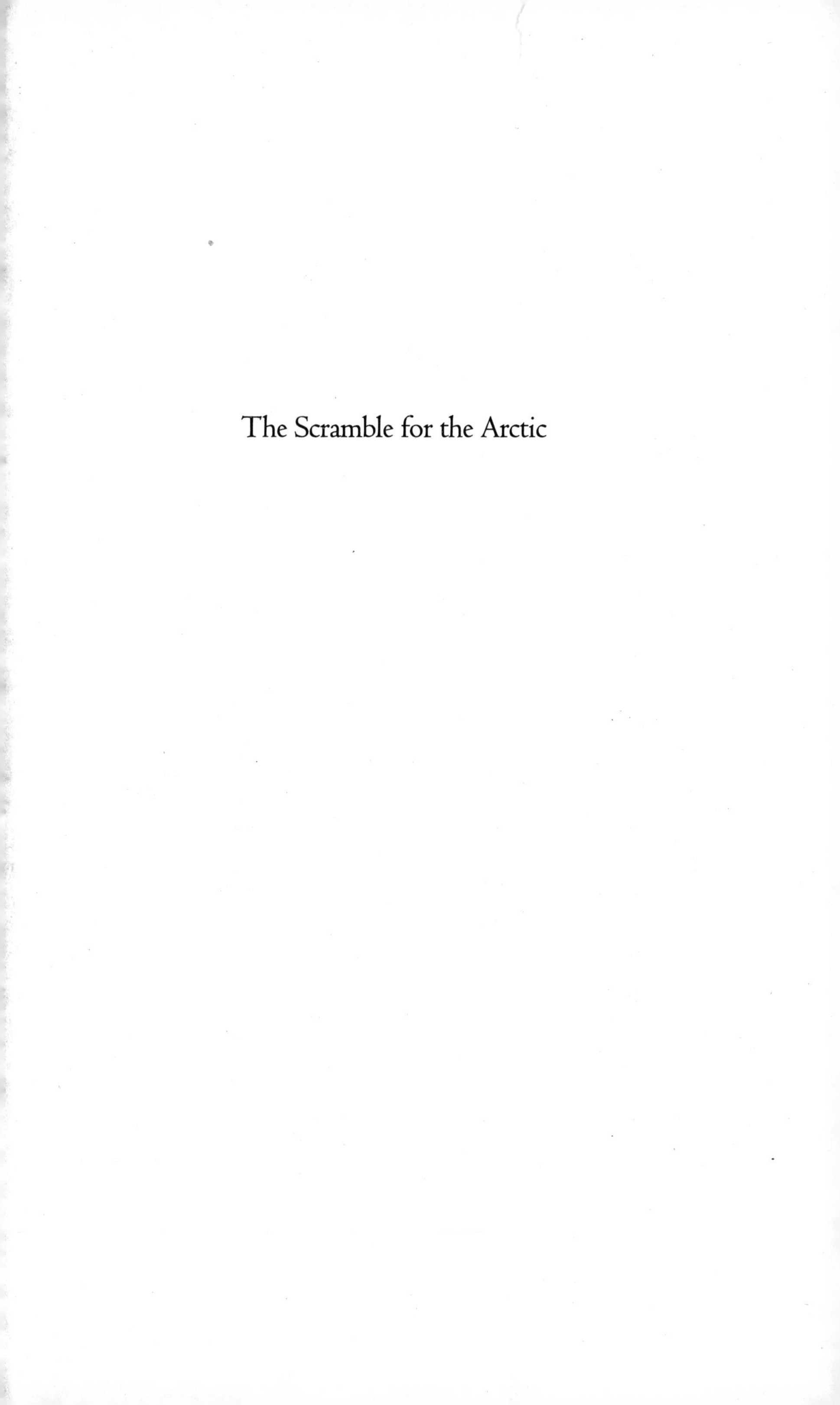

The Scramble for the Arctic

THE SCRAMBLE *for* THE ARCTIC

OWNERSHIP, EXPLOITATION AND CONFLICT IN THE FAR NORTH

RICHARD SALE
EUGENE POTAPOV

F

FRANCES LINCOLN LIMITED
PUBLISHERS

Frances Lincoln Ltd
4 Torriano Mews
Torriano Avenue
London NW5 2RZ
www.franceslincoln.com

The Scramble for the Arctic

First Frances Lincoln edition 2010

A catalogue record for this book is available from the British Library

ISBN 978-0-7112-3040-8

Printed and bound in China

9 8 7 6 5 4 3 2 1

Contents

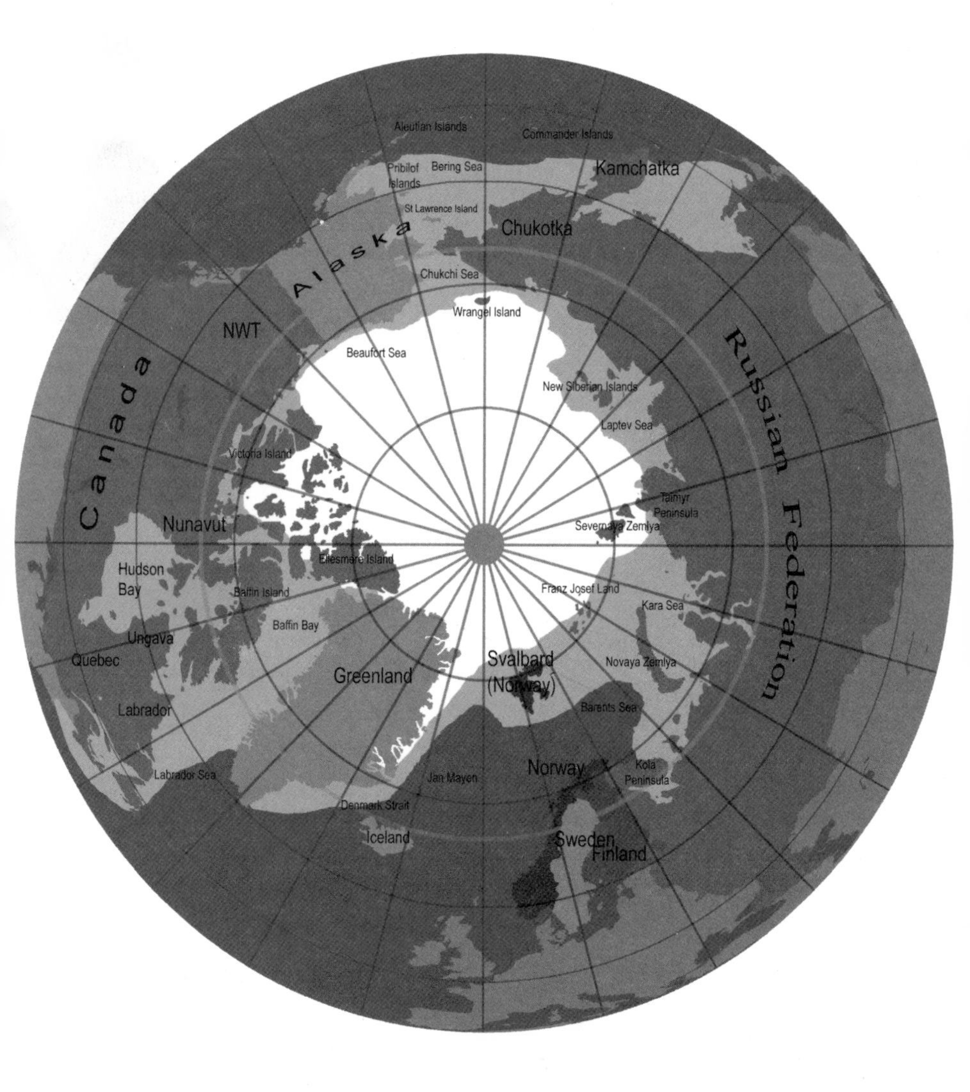
Aleutian Islands
Commander Islands
Pribilof Islands
Bering Sea
Kamchatka
St Lawrence Island
Alaska
Chukotka
Chukchi Sea
Wrangel Island
NWT
Canada
Beaufort Sea
Russian Federation
New Siberian Islands
Laptev Sea
Victoria Island
Taimyr Peninsula
Nunavut
Severnaya Zemlya
Ellesmere Island
Hudson Bay
Baffin Island
Franz Josef Land
Kara Sea
Baffin Bay
Ungava
Quebec
Greenland
Svalbard (Norway)
Novaya Zemlya
Labrador
Barents Sea
Labrador Sea
Jan Mayen
Norway
Kola Peninsula
Denmark Strait
Iceland
Sweden
Finland

Introduction

In 1998 the Inuit Circumpolar Conference made a Declaration on the rights of Inuit peoples to hunt and trade Arctic sea mammals. The Conference had held its inaugural meeting in 1977 with the intention of representing the native peoples of the High Arctic – the sea mammal hunting peoples of Alaska, Canada and Russia. It therefore covers all Arctic native peoples except the reindeer herding peoples of Eurasia – the Sámi of Fennoscandia and the peoples of northern Russia apart from those of the Chukotka coast. The 1998 Declaration noted the importance of sea mammal hunting to Inuit culture and way of life, affirmed the Inuit's right to continue hunting and trading the animals and called upon national governments to recognize these facts and to remove hindrances to hunting and trade. The Declaration emphasized that no government had the right to restrict or regulate the Inuit's sustainable use of the animals.

The Declaration also made a broader point (though one rooted in the Inuit rights that were its main concern) that 'the future of human survival depends upon developing sustainable relationships with renewable resources'.

The timing of the Declaration was ironic, the Arctic sea mammals having been saved from an extinction driven by the relentless overhunting of sub-Arctic nations only for their future to be threatened by climate change brought about, largely speaking, by the same nations.

When the explorers of medieval Europe headed north it was to search for trade routes, to add new lands to national empires, or to look for profitable resources. The history of the exploitation of Arctic resources is one of a complete lack of management, the exploited animals – whales killed for their oil, sable and other animals for their fur – being gathered as long as it was economic to do so. Only when numbers were too low to warrant the expense of the kill was hunting stopped. At the start of the twentieth century that first stage of exploitation was essentially complete. The Arctic was then largely ignored – until the world began to run out of easily exploitable resources, metals and, increasingly, oil and gas.

The exploitation of mineral wealth in the Arctic brings particular problems. Those involved in the work must contend with a harsh climate that tests both men and machines. But the climate also tests the Arctic environment. All environments are fragile: they have evolved slowly and rapid change threatens their ecology. But the Arctic is especially vulnerable to rapid change because of the harsh climate. The growing season for plants, and the breeding season of the majority of animals, is short. In temperate and tropical zones environmental damage may be countered by a relatively rapid recolonization by plants and animals. In the Arctic such damage may take decades or centuries to repair as the speed of the succession is low. Industrial development causes particular problems not only because of the degree of land destruction, but because pollution makes regeneration even lengthier and there is little incentive for companies to clean up: out of sight is out of mind and the Arctic is, in the eyes of many, just a useless wilderness area. In the Russian north reindeer pastoralism, the way of life of the native peoples, has not been possible on formerly good pasture after the cessation of industrial activity.

Exploitation of the mineral wealth of the Arctic therefore threatens northern peoples by damaging their environment. There are many who will argue that these peoples, determined to live in a primitive state, clinging to the traditions of a backward past, should embrace the benefits of urban-industrial society. The changes in the lifestyles of northern peoples wrought by the communist regime in Russia and the liberal democracies of the Nearctic and Scandinavia were justified on the basis that they were for the peoples' own good. The validity of their lifestyles and choices has now achieved (somewhat grudging) acceptance, though each of the Arctic nations has retained the right to exploit the mineral wealth that might exist beneath the ground which the peoples have reluctantly ceded. And the problem that all northern peoples have is that they occupy an area of massive potential mineral wealth.

During the first phase of exploitation of Arctic resources there were no regulations to protect the environment. It might be imagined that the development of international law would now offer protection, since that development has followed three specific principles: that countries should conserve the environment and wildlife; that countries should assess the potential effects on the environment of any course of action and should monitor the environment to obtain information on the actual impact; and that countries should cooperate to conserve the environment outside their own national boundaries. These principles, once enshrined, should aid environmental conservation. What we are now discovering, though it comes as little surprise, is that when a government is faced with a choice between the welfare or economic development of its population and the environment, the latter will invariably come off second best. And since, all too often, governments act in the interests of their own survival, or in the interests of particular groups, the environment is often sacrificed to serve the interests of minorities.

Confucius said that to define the future we must study the past. This book explores the way in which ownership and exploitation proceeded in the Arctic, and the conflicts each brought. It then looks at the position of the Arctic in international law. Finally it considers the impact the exploitation of the area's mineral resources will have, noting that the mining of Arctic oil and gas will further exacerbate the dire problem of climate change that is already faced by the Arctic, its peoples and wildlife. Rising temperatures and diminishing ice cover will destroy cultures and wildlife that are unique. The former is probably inevitable, as preservation would require a theme park Arctic which may be neither desirable for, nor acceptable to, new generations of Arctic dwellers. Loss of the ice will mean extinction for many species, including one of the world's most iconic animals, the polar bear. But perhaps most importantly, the scramble for the Arctic's minerals may lead to conflicts that threaten not only iconic animals, but world peace itself.

A wedding in south-east Greenland. The bride and her page are in traditional clothing of embroidered sealskin.

Chapter One
Human Settlement of the Arctic

For philosophers to interest themselves in speculating on a horde so small, and so secluded, occupying so apparently hopeless a country, so barren, so wild, and so repulsive; and yet enjoying the most perfect vigour, the most well-fed health, and all else that here constitutes, not merely wealth, but the opulence of luxury; since they were as amply furnished with provisions, as with every other thing that could be necessary to their wants.
John Ross, 1830, after surviving his first winter in the Arctic

Eighteen thousand years ago the northern hemisphere was in the depths of the last ice age. The ice had reached its maximum extent, covering north America as far south as St Louis, though, amazingly, leaving much of northern Alaska – that most Arctic of the United States – ice-free. All of Canada, both the mainland and the northern islands, was covered by ice, as was Greenland. Iceland lay beneath an ice sheet, as did most of northern Europe, the ice reaching as far south as Belgium and central Germany. In Britain, ice shrouded mid- and north Wales, and everything north of a line from Birmingham to the Humber. Yet in the Old World, as in the New, there were curiosities in the ice coverage. Although Scandinavia lay under many metres of ice, as did the bordering area of Russia to the east, the northern area of Russia – like Alaska a land renowned for near-Arctic conditions – was increasingly ice-free. Wrangel Island, today an important maternity denning area for female polar bears, escaped the ice, as did much of the northern coast of Chukotka. South of this ice-free area, permanent ice meandered its way across much of Siberia.

Then the temperature rose and the ice retreated. As it did, seeds blown north by prevailing winds allowed plant life to recolonize the newly exposed, barren but fertile land. The plants were followed by the animals, and by the humans of southern Europe who depended on them for food and clothing.

The southern Caucasus were occupied by man around 1.8 million years ago, while recent discoveries in Spain have pushed the occupation of western Europe back to about one million years ago. It now seems certain that they had reached eastern Britain at least 700,000 years ago. Perhaps humans moved even further north and evidence of their occupation is waiting to be discovered. Perhaps it was destroyed by the relentless ice as it moved south.

Ancestral man's occupation of Europe corresponds to the time of the ice age events of the Quaternary period of geological time. During that period the temperature of the Earth was usually a little cooler than today, with occasional periods when it was warmer, though not dramatically so. The Arctic would have been little more hospitable than it is today, and often much less so, and for people with limited tools would probably not have been an inviting prospect. Yet recent discoveries of flint tools close to Russia's Yana River, at 70°N, about 500km north of the Arctic Circle, suggest that man had penetrated deep into the sub-Arctic before the last ice age, the finds being dated to about 30,000 years BP.

After the last glacial maximum (at 18,000 years BP) ice cores taken from the Greenland ice sheet indicate that the temperature of the Earth rose, slowly at first, but then more quickly to a maximum at about 13,000 years BP. It then cooled abruptly, staying cool for 1,500 years before rising sharply. This cooling and rapid warming is known as the Younger Dryas event (named after the spread of the Arctic plant *Dryas octopetala*) and may be significant in terms of the future climate of northern Europe, as it is conjectured that it resulted from a switching off, and then on again, of the North Atlantic Drift, the warm ocean current which elbows the Arctic northwards in north-west Europe.

At about the time the rising temperature reached its maximum, finds at a site close to the Berelekh River in Yakutia, Russia, close to the mouth of the Indigirka River, at 71°N, indicate that, in Eurasia, Paleolithic man had reached the northern mainland coast. It is conjectured that a major prey species of these ancestral hunters was the mammoth, the significant finds of the animals having led to the Berelekh site being called a 'mammoth cemetery'. The finds include tools and weapons of worked stone and mammoth ivory, but also include a drilled pendant of jadeite, indicating an artistic life as well as technical expertise. At the Bluefish Caves, in the northern Yukon of Canada, tools and weapons of stone and bone, together with the remains of butchered animals, are considered to be contemporary with the Berelekh finds. Some archaeologists consider the finds, and those from a sand bar of the nearby Old Crow River, to be even older, perhaps dating to 25,000 years BP, but that view is not mainstream. Further south, and across the border in Alaska, the Broken Mammoth site in the Tanana valley has revealed a vast number of artefacts, including stone tools and the remains of animals such as bison, which have been radiocarbon dated to about 12,000 years BP.

In the 1980s archaeologists found a human settlement on Zhokhova Island, one of the DeLong Islands in the East Siberian Sea, at 76°N. The campsite contained the remains of a dog sled of a very similar design to the ones used by modern Chukchis, stone tools, kitchenware, a knife made of mammoth ivory and an ivory stick thought to have been used to punch holes in the ice, all dated to about 8,000 years BP. Food remains at the camp included polar bear, walrus, seal and reindeer.

Native Peoples of the Nearctic

The age of Bluefish/Old Crow/Broken Mammoth finds is critical to an understanding of human arrival in North America. The recent extraction of DNA from a clump of hair found at Disko Bay on Greenland's west coast and dated to 3,000 years BP has confirmed the Asiatic ancestry of the Arctic dwellers of that time. It is generally accepted that the ancestors of these early dwellers arrived by migration across the land bridge which existed between Chukotka and Alaska at a time when the water locked up in land ice caused a lowering of planetary sea levels and the disappearance of much of the Bering Sea. The exact environment and ecology of Beringia, as the area covered by eastern Chukotka, the land bridge and western Alaska is now known, is still debated. But whether Beringia was a lush land or rather more sparsely vegetated, it is clear that adequate resources existed to sustain the herds of animals which crossed from Asia to America – these including mammoth, Cherskiy horse, saiga antelope and several species of bison – and persuaded human hunters to follow.

The Berelekh and American finds are comparable in age, and to an extent style, with the Clovis culture of North America, which is believed to have been the forerunner of the continent's native Indian peoples. There is, however, much to debate on this issue, not least the fact that finds in South America and the southern United States pre-date (some authorities would prefer the phrase 'appear to pre-date') the Clovis culture, implying an earlier migration. That is not contradicted by studies of Beringia, which imply that the Asia–America land bridge might have been open for much, perhaps all, of the period from 25,000–12,000 years BP. Beringia would also have existed during previous Quaternary ice ages. The finds of *c.*13,000 years BP coincide, approximately, with the known extinction of some thirty-five large mammal species in North America, all of which disappeared over a period of about 3,000 years. The idea that a new wave of hunters crossed Beringia into the New World bringing sophisticated techniques that overwhelmed the existing populations was convenient and seemed plausible. But doubts have been raised that make the idea now seem less acceptable, and most experts suggest a combination of climatic change, the rise of the Clovis culture (which of itself does not require a new migration of hunters), new diseases imported by migrating Asiatic species, and perhaps even as yet undetected events, to explain the extinction.

If the evidence for a more ancient crossing of Beringia is sound, then the people who left the evidence of their existence at the Bluefish Caves and Tanana River may have been a second wave of arrivals, an idea now favoured by many authorities, who see an early migration giving rise to the native peoples of southern Canada and the Lower 48 States, with ancestral Arctic dwellers – the ancestors of the Inuit, Aleut and Athabaskan/Dene peoples – arriving later, perhaps even post-dating the Bluefish/Tanana finds.

Before the idea of successive waves of migration had been mooted, it was assumed that the peoples of Arctic North America had probably originated below the treeline, migrating north and learning new skills that evolved into the distinctive groups which archaeologists of the early twentieth century recognized – sea mammal hunters of the far north and Greenland, caribou hunters and fishermen of the Canadian Barren Lands, and groups which mixed the two, depending on season or circumstance. This idea was challenged by the Fifth Thule Expedition of 1921–4, led by explorer and anthropologist Knud Rasmussen.

Rasmussen was born in Ilulissat (then called Jacobshavn) in west Greenland, the son of the town's Danish pastor and a Greenlandic mother. The Fifth Thule was a remarkable achievement, Rasmussen's team travelling from Thule in north-west Greenland to Nome on Alaska's western coast. The journey included the first dog-sled crossing of the North-West Passage, and would have continued across the Bering Strait to Russia had visas not been denied. The expedition's aim was to solve 'the great primary problem of the origin of the Eskimo race.' While use of the term 'Eskimo' is now considered derogatory over much of North America, the aim, though ambitious, was largely realized. What Therkel Mathiassen, the senior archaeologist on Rasmussen's team, discovered was consistent with the existence of a pan-North American/Greenland Arctic culture – named the Thule culture because the first evidence of its existence was discovered at Thule – dating from around AD 1000. The evidence suggested that the Thule culture was the forerunner of the present-day Inuit population. But though this was a satisfying discovery, it left a gap of many thousands of years in the development of the Inuit. The ancestral peoples were named Palaeo-Eskimo, a name which offered the idea of an understood history, but which merely masked a lack of any true understanding.

Filling the gap was not easy. In one sense the Arctic is generous to archaeologists. The cold preserves artefacts, while the lack of any activity which might disturb ancient sites – for instance agriculture or forestry in southern lands – allows them to remain wherever they were dropped. Change is also slow in the Arctic, the land being covered with snow for much of the year and the growth of vegetation or surface erosion limited after the spring melt. Ironically, the very fact that vegetation is sparse in the High Arctic aids the archaeologist to discover sites, plants seizing on any opportunity to acquire nutrients. A bone leaching nutrients into the soil becomes the centre of a small oasis, the relatively luxuriant growth pointing the way to an ancient site. The unchanging nature of the Arctic helps the archaeologist to find artefacts, and means he has little difficulty in visualizing the landscape in which they were deposited.

But the Arctic is also a less generous workplace: it is a hostile environment for humans not born there to survive its rigours; it is vast, the limited number of sites thinly spread, and travel in any season is difficult. These problems restricted study until after the 1939–45 war. The war saw the development of airbases in Greenland and other parts of the far north, and the development of navigational aids which allowed an expansion of flying in the Arctic. When peace returned, the use of aircraft offered ready access to remote sites and field trips to them did not require arduous journeys and the transport of months of supply.

The first find of the new era of Arctic archaeology was at Cape Denbigh, south of Alaska's Seward Peninsula. There, when the remains of a pre-Thule people dwelling were excavated, chert and obsidian flints of an even earlier culture were found beneath. At the same time, at Independence Fjord in northern Greenland, the flint evidence of another very early culture (named for the fjord) was discovered. When the objects from these two ends of the Nearctic were compared, they suggested a similarity not only with each other, but with those from Stone Age sites in northern Eurasia. There was also a striking similarity between the dwellings excavated at Independence Fjord and those of the Sámi people of northern Scandinavia. Although there was a continuity in the technology of people across the Nearctic, what was equally striking was that in the eastern Arctic the people had lived much further north than any later culture, sites having been found and excavated near Lake Hazen on Ellesmere Island and in Greenland's Pearyland. In the 5,000 or so years since Independence Culture people lived, no one has lived at such high latitudes. Even more astonishing, these people of the High Arctic lived a semi-settled life based on the hunting of muskoxen. Herds of these powerful animals – shaggy and prehistoric-looking, though in fact of much more recent evolution – still roam the High Arctic. Unlike some other herd animals, muskoxen are only semi-nomadic, moving around a relatively small territory in search of the sparse browse that is the only available food. In spring and summer they are in the valley bottoms searching for new growth, in winter on higher ridges where forage exposed by the wind compensates for the chill of the wind itself. Muskoxen provided food and clothing for the Independence people, but the animals use fur rather than blubber as an adaptation against the cold and so did not provide the oil (obtained from sea mammals) that later Arctic dwellers utilized for heat and light. The Independence people also dwelt in tents all year round, rather than moving into the semi-underground houses that were constructed by their successors. The tents were of muskox skins over a driftwood frame, the skins held down by a ring of stones, the rings still marking the position of the camp sites and so allowing them to be located. In the tents cooking and the only available heating would have been from gathered twigs of Arctic willow (the primary food of muskoxen) supplemented by the occasional find of driftwood. In the High Arctic, with its weeks of continuous darkness, hunting perhaps only possible when the Aurora Borealis shed its fragile, shimmering, ethereal light, survival must have been hard-won. Indeed, it is difficult to imagine that any tribe of humans has had a more precarious existence. The Arctic climate was warmer, which probably meant that muskoxen were more abundant than now, but as with all herd animals, muskoxen are prone to the rapid spread of disease. Most herds are small and so could be eliminated entirely. They could also move away if browse became scarce through overgrazing or a bad spring. A shortage of muskoxen or willow twigs for fuel, a lack of driftwood or a lead hunter's disabling injury, and a family or small group of Independence people could have perished in a few months. It is no surprise that in places in the High Arctic, finds suggest that areas occupied for some time were suddenly unoccupied for several centuries.

It is conjectured that the peoples of the High Arctic followed the muskoxen along the northern coast of the Nearctic mainland, then north to Ellesmere and other Canadian islands and across to north Greenland. When they reached northern Baffin Island, some turned south, spreading through the island, then west across Hudson Bay and south to the mainland, settling Ungava. They were Earth's last native pioneers, no other ecosystem on the planet having been settled later. Muskox meat could have been supplemented by fish and, perhaps, the occasional seal at the coast, but the scarcity of animals meant that the population density was low: in the High Arctic it is likely that there were never more than a few hundred humans in every million square kilometres of land.

As the muskox hunters moved east, other groups from the Bering Sea coast were exploring more southerly areas of the mainland. On the tundra there were vast herds of caribou, whose annual migration in search of calving grounds and winter fodder gave people the chance of both a semi-nomadic and a settled lifestyle. Below the timberline, woodland caribou and other animals offered a more secure existence than that of their northern neighbours. These areas were settled by groups who would become the various Athabaskan tribes of central Alaska and western Canada, the Tlingit and Haida of south-east Alaska, and the Algonkian tribes of central and northern Canada.

Apart from the disasters wrought by the precarious nature of their existence, the life of these early Arctic dwellers seems to have been relatively stable for several thousand years. Then the Arctic became colder. In the High Arctic the change was enough to end a lifestyle which was not taken up when the climate warmed again. In the south the treeline retreated southwards, taking the 'Indian' cultures with it, but expanding the range of the true Arctic dwellers who became concentrated on the coastal mainland and around Hudson Bay, though there is evidence that a muskox-based culture still existed in places such as Banks Island where the number of animals remained high. Probably as a consequence of the deteriorating climate the people began to build winter huts. These were partially subterranean, and so had the advantage of presenting a smaller surface area to winter's winds. A covering of sea mammal skins was thrown over a framework of whalebones or driftwood. The edges of the skins were held down with boulders, and the whole was covered with turf and a blanket of snow for insulation. The living area was reached by a passageway that was lower to act as a cold air trap. The hut was heated by soapstone lamps fuelled by the blubber of sea mammals, the same oil providing power for cooking. To the modern eye the lifestyle seems very far from ideal, the environment equally far from hospitable, but though still subsistence hunters operating in a dangerous world, the people had the time to develop a new culture, implying a degree of stability and security that had likely eluded their High Arctic ancestors most of the time. It is difficult to accurately assess the population of the Arctic at this time of transition, but based on collected evidence most authorities now believe that from northern Alaska through to the Atlantic coast of Canada there were probably no more than 5,000 people in total, most living in groups of around 200–300, these fragmenting into family groups, then rejoining as resources required or allowed.

The oral tradition of the Inuit includes stories from their immediate ancestors, the Thule people, which tell of them meeting the 'Tunit' when they reached the central and eastern Arctic. It is believed that the Tunit can only have been the people whose culture developed as the Arctic became colder: from the site of the discovery of the first artefacts which introduced this new cultural tradition to archaeology, Cape Dorset on south-west Baffin Island, these new Arctic dwellers are known as the Dorset Culture. The Thule were astonished by the Tunit. They were taller than the Thule, and immensely strong, capable of building with huge boulders. They were a peaceful people whose skill in producing tools and weapons, and in carving beautiful art, impressed the Thule. But in contrast to the newcomers, they were a very primitive people. They had few dogs, despite their ancestors having made use of them to corner muskoxen. They had few sleds and if the dogs they had were used to haul them, the attachments had none of the sophistication of the Thule. The Dorset people had abandoned the bow and arrow, hunting sea mammals – seal, walrus and, occasionally, beluga and narwhal – with harpoons, and land mammals with spears. As it now believed that bow and arrow technology was probably transferred from Palaeo-Eskimos to the native Indians to the south, it is ironic that it was lost by the Dorset people.

The Dorset had also abandoned the kayak. Hunting from a kayak requires floats to tire the prey and to keep it from diving when struck, and to prevent dead animals sinking. It is a highly skilled and dangerous occupation. The Dorset used the less risky, but less efficient, technique of hunting from the floe edge (the edge of the sea ice). Consideration of the effort required to haul a struck or dead walrus or whale from the sea is a vivid illustration of the immense strength the Thule saw in the Dorset.

But despite the limitations of their tools and techniques the Dorset were able to spread across the Arctic, resettling Greenland which had been abandoned by their ancestors. The colder climate meant both a greater abundance of sea ice and its greater persistence, which suited the Dorset as sea ice was the platform from which they hunted. For the thousand years that the Arctic was colder the Dorset Culture dominated the Arctic east of the Coppermine River. Then the climate changed again, becoming warmer. At first the Dorset people were able to adjust, moving north in search of colder areas where their sea ice hunting techniques were still appropriate and abandoning southern sites, but then, in about AD 1000, a new wave of settlers arrived from the west to displace them.

While the pre-history of the central and eastern Arctic are well understood, that of the western Arctic is much less so. Alaska, the Aleutian Islands and Chukotka border a sea of immense productivity and there is little hindrance to the transfer of ideas and techniques from Eurasia which the thousands of kilometres of ice and tundra to the east represented. The development of the area is correspondingly complex; several cultures have been identified, but the relationship between them, and the overlap of them, cause problems for the archaeologist or anthropologist intent on producing a credible timeline. What is clear is that ancestral peoples, beginning with the somewhat prosaically named Old Bering Sea Culture, developed the ability to hunt bowhead whales. The whales, by far the largest mammal to inhabit the Arctic, migrated annually past headlands where posted look-outs could observe them. While killing a whale was difficult and dangerous, the sheer quantity of food produced allowed large, permanent settlements to develop: one bowhead could keep an entire community well fed throughout the winter. During the last centuries of the first millennium AD the Bering Sea cultures developed and imported a series of innovations which were to lead to the Thule Culture dominating the North American Arctic. They had fast, highly manoeuverable kayaks from which they hunted the bowhead. They also had a larger boat, the umiak, or women's boat, which could hold up to twenty people. They had dogs and efficient ways of attaching them to sleds. They had a recurved bow and plate armour, technology imported from the Mongolian steppe. They also had iron.

As far back as the later Palaeo-Eskimo peoples in the eastern Arctic there had been a limited amount of iron, sourced from the three large meteorites found close to Cape York in Greenland (whose later removal was one of several unsavoury escapades in the career of Robert Peary). There was no smelting or true metal industry; flakes of the meteorites – which had a high nickel content making them hard to work – were simply knocked off and then hammered to create an edge that was long-lasting and could be reworked. It is known that Cape York flakes were traded in the eastern Arctic. Copper was also worked, deposits close to the mouth of the Coppermine River being known from pre-Dorset times, though working seems to have been minimal. It has been suggested that copper working only really began in earnest after the use of meteoric iron became established. The softer, but more easily worked, metal was used by peoples who knew of, but had no easy access to, iron.

Perhaps news of the meteoric iron, and even of the copper, reached the Bering Sea, and perhaps the Thule decided that heading east offered an easier option for acquiring iron than trading with the more advanced peoples of Eurasia. Tantalizingly, the eastern expansion of the Thule corresponds, approximately, with the arrival of the iron-using Norse in south-western Greenland. It is known that Norse iron was traded as far as Baffin and Ellesmere islands. Did news of it also reach the Thule?

Whether it was a search for iron, an escape from conflict around the Bering Sea (the possession of armour suggests the Thule had experience of warfare), or merely an expanding population seeking new lands, from about AD 1000 the Thule left northern Alaska and travelled east. Beyond the Coppermine River they would have met the Dorset/Tunit. There is some evidence to suggest that the two cultures shared living spaces for a while, though the oral tradition of the Inuit maintains that the Tunit fled their homes. It is possible that the Thule, smaller but better armed and more aggressive, put the original owners of the land to flight, but the complete disappearance of the Tunit would require a genocide which does not form part of oral tradition. More likely is that diseases brought by the newcomers rampaged through a population with no immunity to them. The survivors, driven from their homes by fear, now isolated and helpless, died of starvation and despair. And in the space of just a few hundred years the Thule became masters of the Arctic from Chukotka to Greenland.

As noted earlier, the Thule are believed to be the ancestors of the present-day Inuit, the peoples who inhabit the central and eastern Arctic. The name 'Inuit' is itself modern, and means 'the people' in Inuktitut, the language of the Canadian Arctic dwellers: a single person is an Inuk. The name change represented a reaction against the name 'Eskimo'. The latter is said to derive from the Athabaskan for 'raw meat eater', though it is worth noting that there is not universal acceptance of this derivation. Some authorities point to an Algonkian source deriving from 'snow-shoe hunter', while others suggest an alternative Algonkian meaning, deriving only from 'speaker of a foreign language'. Since the Thule people did indeed eat raw meat – it was the source of their Vitamin C – that derivation has merit, while the use of snow-shoes was more confined to forest dwellers, where snow can drift into soft heaps, rather than the tundra, where the wind tends to form a hard crust known as sastrugi. The Inuit did, of course, speak a language foreign to the forest dwellers of the south, but the use of 'Eskimo' by a people to the south whose relationship with the northern dwellers was often unfriendly, was seen as derogatory and was eventually the cause of resentment. In Greenland, the people call themselves the Kalaallit (singular Kalaaleq), which means 'the Greenlanders' and so is synonymous with 'Inuit'. In the Greenlandic language (Kalaallisut) Greenland is Kalaallit Nunaat, 'the land of the Greenlanders'. Linguistically the Inuit and Greenlandic languages are linked, but differ from the basic language of the peoples of the western Arctic. There, the name 'Eskimo' is also considered much less derogatory, and it is not unusual to hear a native referring to himself as a Yupik Eskimo. 'Yupik' is the singular form of Yuppiat (occasionally rendered Inupiat) for the peoples of northern and western Alaska, St Lawrence Island and the neighbouring Chukotka coast. To the south the people of Kodiak Island, the Alaska and Kenai peninsulas and around Prince William Sound are the Alutiit (singular Alutiiq).

The evolution of the peoples of the Bering Sea – those given above and the Aleuts of the Aleutian Chain – is complex and still subject to debate. Indeed, the distinction of the northern and western populations of Alaska into just Yuppiat and Alutiit would be challenged by some authorities. It is now believed that the Aleuts (the Unangan – 'coastal people' – in their own language) have occupied the island chain linking south-western Alaska to the Commander Island and Kamchatka for at least 7,000 years, perhaps longer. (Some authorities believe that ancestral Aleuts may have moved into the western island chain as early as 10,000 years ago.) In contrast to the Palaeo-Eskimo societies, and their successors, Aleut society was hierarchical, with a hereditary noble class, commoners and slaves, the latter often captured during periods of warfare. Wars might be between islands, but there were also conflicts with both the Alutiit to the east and the Yuppiat to the north. Aleut society was sedentary, and large villages, probably the largest to have been seen in the pre-contact Arctic, grew up in favourable harbours. Aleut houses (*barabaras*) were of unusual form, being semi-subterranean, with a framework of driftwood (there were no trees on the Aleutians) covered in turf and entered by a hole in the roof. The Aleut hunted sea mammals, particularly the Steller's sea lion, and whales, but also hunted caribou and fished for salmon in spawning rivers. Sea mammals were hunted in *baidarkas*, a form of kayak unique to the Aleut, with a bifurcated bow like the open jaws of a fish. The boats were occasionally paddled by two men. Aleuts sometimes hunted whales using a poison-tipped harpoon, the poison deriving from the Kamchatka aconite (*Aconitum maximum*). As the poison was slow-acting in massive whales, the stricken animal had either to be watched for hours or days, or to be (hopefully) retrieved when it washed ashore.

At the time of first contact, eastern Arctic dwellers occupied the Aleutian island chain, the Alaskan peninsula, Kodiak Island and the nearby coast of mainland Alaska, the western and northern coasts of Alaska, and a strip of northern Canada which was narrow near the McKenzie delta, then broadened, reaching Hudson Bay almost as far south as the Churchill River. On the eastern side of the Bay, the Inuit occupied the Ungava Peninsula, the Hudson Bay coast as far south as Cape Henrietta Maria, and much of Labrador. To the south were the tribes of Athabaskan and Algonkian Indians. Of the former, one of the most important was the Gwich'in (or Kutchin) of north-east Alaska and Canada's Yukon Territory. Their life was based entirely on caribou, hunting them twice as they passed on their annual migrations to and from the calving grounds. Caribou meat was cached for the periods when the animals were out of range of hunting parties, the otherwise monotonous diet supplemented by fish taken from, for instance, the Porcupine River, as well as various birds. To the east, the Chipewyan were also caribou hunters. The conflict between Chipewyan and the Inuit is noted by the Hudson's Bay Company employee Samuel Hearne who made a famous journey in search of a river which, legend had it, flowed between banks of solid copper. In his book of his journey to the mouth of the river – named the Coppermine – he describes how his Chipewyan guides surprised an encampment of Inuit and slaughtered them all before his eyes. One particularly terrible moment occurred when an Inuit girl, about eighteen years old, already wounded by spears, clung to Hearne's legs and, despite his entreaties, was speared again, the Indians asking the Englishman if he wanted an Eskimo wife. The savagery of the massacre, and the contemptuous way in which Hearne's pleadings were treated suggest that relationships between the two native groups were invariably hostile. Of the Algonkian tribes, the most important were the Cree whose assistance in the European opening up of Canada will be considered below.

In Greenland, in general, Inuit/Greenlanders inhabited the west coast southward from Upernavik, and the more rugged east coast north to about Angmassalik. 'In general' because the nomadic Inuit might actually be found anywhere, though much less likely and in much smaller numbers, than in the more hospitable southern areas. Since the Thule almost certainly arrived in Greenland from Ellesmere Island, they might have gone either south, or around the northern coast to the eastern side of the island: on one Royal Navy expedition in 1823 a ship commanded by Douglas Clavering found a group of twelve Inuit on the island now named for him off Greenland's north-east coast. It is not clear whether they were heading north or south, but they were certainly a long way from the 'inhabited' section of the east coast. There were also Inuit around Thule in the north-west of the island. Between them and their southern cousins was a section of coast completely ice-covered, Greenland's ice sheet reaching the sea along its entire length. The long, dangerous sea voyage required to bypass this ice is the probable reason for the continuation of the myth that the people, the so-called Polar Eskimo of Thule, believed themselves to be the only inhabitants of Earth, and that the rest of the planet was covered in ice. This myth arose after they were first contacted by a ship of the Royal Navy on its way to search for the North-West Passage, a contact memorable for many reasons.

Native Peoples of the Palearctic

In 1799 Martin Wahl, Professor of Botany at the University of Copenhagen, was collecting flower specimens close to Tromsø, the first person to do so in any systematic way. As he made his way through a forested area close to Balsfjord he chanced upon an exposed boulder. To his amazement he noticed that a 'figure of a buck is hewn in the stone surface'. From the drawing in his notebook the buck would appear to be an elk (the Eurasian 'elk' is *Alces alces,* the 'moose' of the Nearctic: the occasional confusion caused by different names for the same animal is amplified because in the Nearctic there is also an 'elk' – *Cervus canadensis* – also called the wapiti). Wahl had discovered the first example of an ancient petroglyph (rock art) in northern Fennoscandia. Today dozens of petroglyph sites are known, spread along the Atlantic and northern coasts of Norway, in northern Sweden, in Finland, and on the Kola Peninsula and White Sea coast of Russia. The art usually comprises depictions of prey animals – reindeer, elk (moose), birds – most notably swans and loons – fish and other animals, such as bears, which would have been a potential threat to hunters, boats and shamanistic symbols, humans (sometimes with explicit private parts), and occasionally abstract forms. There are curious geometric shapes, often net-like. These are, perhaps, fishing nets, but are also occasionally linked to form fence-like structures. That is likely to be what they were: fences of woven branches being used to herd reindeer into killing areas. The rock art shows three distinct phases: initially the images were polished onto the rock; later they were picked out, perhaps with an antler or a sharp stone; finally they were painted. The images cover a period from the Mesolithic (Middle Stone Age) through to the Bronze or early Iron Age, that is from about 6,000 years BP through to about 2,500 years BP. They therefore post-date the finds at the Berelekh River.

Dating the rock art would seem a near impossible task, but the last ice age offers a method. Ice is heavy, a one metre cube weighing about 1 ton, and the ice sheet which covered northern Fennoscandia was 1,500m/5,000ft thick (perhaps more) in places. The weight of the ice compressed the land below it, and when the ice retreated the land rebounded. This effect – known as isostatic rebound – occurs at a predictable rate: in northern Sweden and Finland the land is rising at a rate of up to 9mm per year (about 1m/3ft per century). For inscribed boulders, many of which are near the sea, rebound therefore allows a good estimate of age, it being assumed that the artist would have been working on dry land.

The petroglyphs of northern Fennoscandia indicate that as the ice retreated and the barren ground left behind was recolonized by plants, birds and animals, early hunters followed, settling the Arctic fringe. Following the sudden cooling, then rapid warming, of the Younger Dryas period (the rapid warming corresponding to the occupation of the Berelekh River and other sites in northern Russia), there were further temperature troughs and peaks which some have conjectured were also associated with interruptions of the North Atlantic Drift. When stability returned, north-western Europe would, then as now, have experienced warm, damp winters (relative to the Nearctic, that is: few visitors to Arctic Fennoscandia in mid-winter are likely to make much use of 'warm' as an adjective when discussing the weather).

Apart from the petroglyphs and a few other scattered finds, little has been discovered about the Stone Age dwellers of Fennoscandia that does not conform to the lifestyle of their better-studied, southern cousins. For a more complete picture of northern Europe's Arctic dwellers we need to move forward in time to the last dates of the rock art, when metal-using peoples were moving north to occupy both Fennoscandia and the tundra of northern Eurasia and, of course, crossing Beringia to people the northern Nearctic.

The settlement of northern Eurasia, particularly north-western Europe, differs from that of North America. There, the Arctic peoples developed essentially in isolation from contact with more sophisticated civilizations to the south. In Eurasia, this was true to an extent for the peoples east of the Urals, but for those to the west there was more or less constant interaction between the Arctic dwellers and the civilizations of southern Europe or southern Asia. In his *Germania*, published in AD 98 the Roman writer Tacitus (AD *c.*56–*c.*117) wrote:

> In wonderful savageness live the nation of the Fennians, and in beastly poverty, destitute of arms, of horses, and of homes; their food, the common herbs; their apparel, skins; their bed, the earth; their only hope in their arrows, which for want of iron they point with bones....

> Nor other shelter have they even for their babes, against the violence of tempests and ravening beasts, than to cover them with the branches of trees twisted together. . . . Such a condition they judge more happy than the painful occupation of cultivating the ground, than the labour of rearing houses, than the agitations of hope and fear attending the defence of their own property or the seizing that of others. Secure against the designs of men, secure against the malignity of the Gods, they have accomplished a thing of infinite difficulty; that to them nothing remains even to be wished.

It is not absolutely clear which people Tacitus was describing. Was it the Finns, or a people who lived in northern Fennoscandia, i.e. the Sámi? The identification is not helped by Tacitus' comment that north of the Fennians there were other, even more primitive peoples. A study of the languages of the present Arctic dwellers of Fennoscandia and eastern Russia suggests that the ancestors of the groups in these two areas were different. The Sámi languages (several exist that are sufficiently different for there to be no identifiable single tongue), together with Finnish and the so-called Samoyedic languages of the native peoples of Russia who reside west of the Taimyr Peninsula, form part of the Uralic language group, which seems to have its origins in the southern Urals. The group also includes Hungarian, suggesting that an ancestral people moved north-west and north-east into the Arctic, but also west towards central Europe where their progress was halted by peoples who had already settled. The languages of most of the Russian native peoples to the east of the Urals are part of the Altaic language group, which has its origins in the central Eurasian steppes. The main exception is the language of the Chukchi and, probably, the peoples of the Kamchatka Peninsula. The latter are usually grouped with the Chukchi, but some authorities consider the native dwellers of the southern Peninsula to speak a language more closely related to the Altaic group. However, the majority of the population of the Peninsula is now Russian. The native population of coastal Chukotka speaks the Yuppiat Eskimo language of the north-west Alaska native people. While this is not surprising in view of the proximity of Chukotka and Alaska, it is unclear whether the Yuppiat evolved in Chukotka and spread east, or evolved in Alaska and resettled north-eastern Eurasia.

The Sámi

The point at which proto-Sámi became ancestral Sámi, i.e. when it becomes clear that the peoples of northern Fennoscandia moved on from being Neolithic hunters to being recognizable as early versions of the people now residing in the area, is still debated, but is now generally assumed to be with the introduction of metal. So, as noted earlier, the distant ancestors of the Sámi were moving into northern Fennoscandia about 2,000 years ago. They occupied both coastal and inland sites, some groups even moving to the larger islands off the north-western and northern coasts. Essentially nomadic, in much the same way as the Inuit were, the Sámi followed the reindeer herds and exploited the annual cycle of fish and sea mammals. The diverse scenery and probable smallness of the Sámi population enabled groups to develop in relative isolation, so that it is possible to identify forest, tundra and coastal Sámi, this allowing the development of the dialects that mean that there is no genuine single Sámi language.

It is likely that only seals and walrus were hunted by coastal Sámi, though smaller whales may have been taken if the opportunity arose. Walrus tusks appear to have been traded with the people to the south, as were furs and falcons. The latter would have been gyrfalcons (*Falco rusticolus*), the world's largest falcon, and, as such, much prized. Such trade goods may have been the reason Tacitus knew of the Finni, and were the likely reason tales of them were also related by the Graeco-Roman historian Procopius (*c.*500–*c.*565). Procopius refers to 'skiing Finni' implying that from an early period the Sámi had acquired the skill. The same expression was also used by the Danish historian Saxo Grammaticus (*c.*1150–1200), though he calls the people 'Lapps'. It was that name by which they were known for many centuries, though now the people's own name for themselves – Sámi – is used. In similar fashion to Inuit, Sámi means 'ourselves' or 'the people': the Sámi land being known as Sápmi ('land of the people').

Evidence for the lifestyle of the Sámi comes from later sources such as the semi-legendary Swedish king Ohthere (sometimes Ottar) who is said to have journeyed to Britain to visit the Anglo-Saxon King Alfred or, perhaps, to have written him a letter. Whatever the exact circumstances, Alfred received information which included news of the *Finnas* which supports the idea that they were hunters and fishermen who traded with the Viking kingdom(s) to the south.

Russia's Native Peoples

To the east of the Sámi, strung out across the northern Russian mainland, are groups of native peoples. First are the Nenets who occupy the coastal area from the eastern end of the White Sea to the western bank of the Yenisey River. That area includes the Yamal Peninsula, and the Nenets also settled Kolguev Island and the southern island of Novaya Zemlya. Ancestral Nenets are believed to have migrated north along the Ob River. Across the Yenisey were the Enets. The similarities of their language to that of the Nenets suggest a common ancestry, but despite this there was considerable antagonism between the two groups. The Enets originally occupied land to the west of the Yenisey as far as the Taz River, and also well to the south. But over time pressure from the Nenets and from aggressive peoples to the south pushed them north and east. Finally in 1849 a battle between Nenets and Enets, the latter in combination with more southerly native groups, stopped Nenet advances eastwards, the two peoples agreeing on the Yenisey as a boundary between them. Once numerous, the Enets are now the most endangered of Russia's native Arctic peoples, with a population of less than 250. Eastwards again, the Nganasan occupied the western coast and central region of the Taimyr Peninsula.

The Nenets, Enets and Nganasan, all sharing a common ancestry and language root, were originally called Samoyed, the derivation of which is debated. Some authorities claim it has the same root as Sámi, pointing to Suomi, the Finnish name for Finland which has the same origin. But others contend that it derives from 'cannibal', bestowed on the peoples by their southern neighbours who either thought them guilty of cannibalism or used to show their contempt for a primitive people. Today, the peoples use the names they call themselves, all of which mean 'people', 'ourselves' or 'man', though Samoyed (or Samodian) is still used to describe the language group of these western Russian natives.

To the east of the Nganasan are groups with a different ancestry (see above). The Dolgan of the south and east Taimyr Peninsula are one of the youngest of Russia's northern peoples, having moved north only in the seventeenth century after breaking away from the Yakuts to the south and joining with the Evenki. The Evenki occupy the vast area between the eastern Taimyr Peninsula and the Kolyma River, though since the 1930s the people of the eastern part of this range have been considered a separate group, the Even. Both peoples have a common ancestry, groups from Manchuria migrating north, and then being pushed further north when the Yakuts pushed north from Mongolia. The Even were pushed north and east by the incomers, some reaching Kamchatka where they may have aided the local people by bringing new ideas and technology.

Across the Kolyma are the Chukchi, who, according to their own legends, came from the south and took over the Yukagir reindeer herds. Yukagirs, the people of the mountains and tundra along the Kolyma River, were no match for the invaders and rapidly declined in numbers in those areas settled by the Chukchi, who have a common ancestry with the peoples of the Kamchatka Peninsula – the Koryaks and Itelmen – and a language that differs from the Altaic group. (Itelmen were often called Kamchadal by the Russians who first settled the Peninsula. Strictly speaking, Kamchadal referred to offspring of relationships between Itelmen and Russian émigrés, but when émigrés outnumbered the native population and such offspring became numerous, the name became interchangeable with the native population itself so that use of Itelmen all but died out.) Finally, coastal Chukotka is occupied by Yuppiat Eskimo, who are related to the people across the Bering Sea in Alaska. In some cases they are truly related, the separation of families, particularly those on the Diomede Islands, being one of the sadder aspects of the Cold War.

At first all the native peoples of northern Eurasia were hunters, living off the vast herds of reindeer that roamed the northern edge of the taiga (the vast forest that blanketed Russia from its European borders to the Pacific Ocean) and the tundra. The people would also have hunted other forest animals, both for food and for furs, and fished the highly productive rivers that drained Russia in to the Arctic Ocean. The Yuppiat of the Chukotka coast hunted sea mammals, including the bowhead whale, just as they did on the Alaska side of the sea, but although it is likely that other peoples took seals and walrus, perhaps even the occasional whale, there does not seem to have been any culture of hunting at the ice edge.

Reindeer herding, as opposed to hunting, is a relatively new lifestyle, despite what is often assumed to be the case. The domestication of the dog was already millennia old when man first decided to farm rather than hunt reindeer. At first domestication seems to have been to acquire a beast of burden rather than a provision of food and skins. Reindeer are a migratory species, always on the move in search of fresh browse, with an annual trek to northern calving grounds timed to coincide with spring growth on the tundra, and a southern journey towards the forest for winter foraging. Hunters living off the herds need to follow this annual pattern, and the first domesticated reindeer helped by hauling tents and the other necessities of life. The use of reindeer for sledge hauling dates to about the sixth century AD, and is believed to have begun among the Evenki/Even. Such a good idea took little time to spread from Fennoscandia to the Kamchatka Peninsula, but a true reindeer pastoralism did not appear for many centuries, perhaps not for a thousand years. Why it took the native peoples that long to engage in what now seems so natural an occupation is still a matter of debate. One of the more interesting ideas involves the biology of the reindeer. In summer the animals must gain weight to counteract the rigours of winter, but that is the time when they are under greatest stress from biting insects, mosquitoes (the bane of life for all Arctic travellers) and the more unpleasant warble and nose bot flies. If the temperature is high – above about 10°C/50°F – the animals suffer further stress, the combination of heat and insects occasionally causing them to cease feeding. Winter mortality then rises. Before the sixteenth century the Earth was in a warm phase and it is conjectured that only when the climate cooled slightly, allowing reindeer numbers to increase, did herding become viable. It is also possible that the domestication of animals by southern peoples, the agricultural communities of eastern central Europe and the steppes, was also a model once contact had been established. The latter reason may explain why the native peoples of the Nearctic, even those who lived almost entirely off them, never domesticated the caribou: the southern 'Indian' tribes of the United States did not domesticate the buffalo either.

The estimated population of true Arctic dwellers in the Nearctic is 60,000, though this is subject to debate, particularly as some estimates of the Eskimo population of Alaska put the figure as high as 40,000 at first contact. The more productive Bering Sea would support a higher population, but even so the Inuit population of Canada and Greenland is likely to have been higher than 20,000, though perhaps no more than twice that figure. The Aleut population at first contact is estimated to have been 15,000–18,000, which makes the Alaskan Eskimo population more acceptable.

In the Palearctic the question of population is more difficult to answer as the effect of Russian expansion into Siberia varied from one group to another. A nineteenth-century Russian ethnographer predicted a gloomy future for all the native peoples east of the Urals, seeing decline and extinction as inevitable. The view was confounded by the fact that while some groups did indeed decline in numbers, others followed the general trend of populations at that time by increasing, in some cases very sharply. The best estimates for pre-contact populations are about 7,500 for Nenet and Enet, with the latter at about 3,000, though they had once been as numerous as the Nenet. This figure is low in comparison to the estimate of 15,000–20,000 Sámi, but the latter is inflated by the inclusion of forest as well as tundra and forest/tundra groups, while the Nenet/Enet figure is for tundra and forest/tundra only. The Yakut and Dolgan population may have been at 6,000 with Nganasan at 2,000. The Evenki/Even group probably comprised 6,000–8,000, with the Chukchis at about 6,000. On the Kamchatka Peninsula there were 15,000 Itelmen in the south where the forests and salmon-stocked rivers offered good living, about 6,000 Koryaks further north where the forest gave way to tundra. These figures must be viewed with caution, but are consistent with the figure for population density calculated on the assumption of a subsistence lifestyle. These give 3 people per 100 square kilometres for the western native peoples (i.e. the Nenet) reducing to about 1 person per 100 square kilometres in inland Chukotka.

First Contacts

It is very likely that the first contact between Europeans and Arctic dwellers occurred in western Fennoscandia when early Norse peoples met the Sámi. The Norse (Vikings) were descendants of Germanic tribes who had moved north into southern Scandinavia around AD 500. Blood testing of the Sámi indicates that those from the western area of Sápmi are more closely related to Europeans than those from the east, supporting the view that the earliest contact was in southern Norway and/or Sweden, perhaps as early as the sixth or seventh centuries. By the second half of the ninth century the Norse had used their amazing seamanship to reach and settle Iceland, and a hundred years later had reached Greenland. In 986 the first Norse settlement was founded in west Greenland. The Norse moved north along the coast and certainly met the Dorset people, probably soon after the first settlement was founded. As the Greenlandic Norse colonies lasted until the early fifteenth century it is likely that the Norse also met, and probably traded with, the Thule people as well.

In Russia the Pomors, ethnic Russians who moved north and settled the White Sea and Kola Peninsula coasts, had met the Nenet by the eleventh century. Whether they met any groups further east seems unlikely. However, the Franciscan monk Giovanni de Piano Carpini (*c.*1180–1252), an emissary of Pope Innocent IV to the Mongol court, did report the existence of peoples living north of Mongol lands. This may only have referred to people such as the Yakuts who occupied the Russian sub-Arctic, but it does suggest the possibility that the Mongols were aware of people living as far north as the tundra.

In the mid-sixteenth century the Russians crossed the Urals and took possession of Siberia, moving east quickly so that within a hundred years they had reached the Bering Sea and made contact with all the native peoples of the Arctic coast. Then in the early eighteenth century, as a result of their search for a possible connection between Asia and America, the Russians reached Kamchatka. They landed on Alaska in 1741. Russian interest in Alaska was largely confined to the south, south-east and Aleutian Islands, making contact with the Aleuts and native 'Indian' populations, but Otto von Kotzebue travelled up the west coast as far as the village that now bears his name and so would have made contact with Yuppiat Eskimos. Almost all Russian contact with Alaskan native peoples were coastal.

In the Canadian Arctic the Englishman Sir Martin Frobisher met Inuit hunters on Baffin Island in 1576, taking an Inuk hostage when some of his men disappeared. The Inuk, and his kayak, were taken back to London where, soon after arriving, the man succumbed to pneumonia. In the following centuries the expansion of the area of interest of the Hudson's Bay Company led to further contact with the Inuit, though the native 'Indian' peoples were the main contacts, while repeated attempts to locate a North-West Passage also resulted in contact. Most famously, the first Royal Naval expedition in search of the Passage, led by John Ross, resulted in contact with the Polar Eskimo of north-west Greenland (and the myth that the Eskimo believed themselves to be the only people on Earth).

The meeting of John Ross' first expedition and the Polar Eskimos of north-east Greenland.

Chapter Two
Ownership

The *Kabloonas* in becoming and respectful terms explained the import of their visit, which by the Esquimaux was answered by a polite invitation to partake of a slice of blubber; the *Kabloonas* inquired how long they were going to remain in their present position; the Esquimaux inquired if they had about them, such a thing as a few fish-hooks; the *Kabloonas* told them they had taken possession of their country in the name of George the Fourth; the Esquimaux told them the seals began to be very scarce.

Robert Huish and William Light, *The Last Voyage of Captain Sir John Ross, R.N. to the Arctic Regions*

This passage highlights the conversation between John Ross, the commanding *kabloona* (as the Inuit called the White Men), and the Inuit over ownership of the land. Though the lack of understanding owed something to unfamiliarity with each other's language, it is also true that the Inuit did not understand the concept of land ownership. So alien was the idea that, for them, the British might as well have been telling them that they now owned the sky. The 'settled' people of Europe had a continuing problem with native Arctic dwellers who followed a nomadic, and therefore alien, lifestyle. Interestingly, the problems that the Scandinavians had with the northern peoples were less pronounced, probably because their concept of *Allemansret* (everyman's right – a freedom to roam) was closer to a nomadic lifestyle.

We start this tour of the Arctic nations approximately mid-way between the Nearctic and the Palearctic, in Iceland, from where we head east.

Iceland

The first-time visitor to Iceland will likely arrive by air, the sea journey to the island being a trip for the connoisseur, that special person who is content to spend long hours gazing across a grey, angry north Atlantic, although the stop at the Faroes – assuming the ship goes that way – is a compensation.

Of course those that arrive by ship follow in the wake of the earliest inhabitants of the island. In terms of the history of human settlement, Iceland has a phenomenal advantage over other countries in having been uninhabited, then settled at a time soon after the development of writing. Two books, the *Íslendingabók*, the Book of the Icelanders, compiled by Ari Thorgilsson, an early-twelfth-century priest, and the *Landnámabók*, the Book of Settlement, which describes the settlement of the island, list the early arrivals and their descendants, together with significant events in the island's history. The *Landnámabók* dates from around the same time as the *Íslendingabók*.

Although the two books were compiled about 250 years after the first settlement and so rely on oral tradition, the detail in the books, which can be corroborated with independent studies, suggests they are very accurate. The books talk of a Norse sailor seeing the island when he became lost on his way to the Faroes, and of a second Norse sailor who circumnavigated Iceland, proving that it was an island. The first settler is claimed to have been Ingólfur Arnarson who was escaping the disturbances caused in Scandinavia by the expansionist ambitions of Haraldur Harfagri (Harald Fairhair) who was to become Harald I of Norway. The books give 874 as the date of Ingólfur's arrival, though archaeology suggests a somewhat earlier date. In true Viking chieftain style, Ingólfur is said to have thrown his carved seat pillars overboard, promising to settle wherever the gods chose to wash them ashore. Legend has it that it took three years to find the pillars in a bay called Reykjavik (Smoky Bay – the smoke being the steam rising from geothermal sites). More likely is that the bay represented the best harbour found by the early settlers. It is also claimed, with some justification, that Ingólfur's first landfall was actually Ingólfshöfði, an island off the southern coast close to the Skaftafell National Park. It can be reached by wading through shallow waters – at low tide the wading is only to boot height – across the glacial outwash plain which now links the island to the mainland. Ingólfshöfði's 100m/330ft cliffs are one of the better seabird nesting sites on this section of coast: a memorial was erected in 1974 to commemorate the 1,000th anniversary of Ingólfur's landfall.

Though the discovery of Iceland appears to have been by chance, it is likely that the existence of a large landmass was inferred by the inhabitants of Britain and Ireland from bird migration and, after the Faroes had been reached, perhaps even by watching weather patterns and cloud formations (the Faroes were reached 200 years before). This is relevant to Iceland's story, as the *Íslendingabók* and *Landnámabók* say that when the Norse arrived, Iceland already had a small population of Irish monks. In AD 825 the Irish monk Dicuil wrote *De mensura orbis terrae*, a summary of the geography of known lands. In it he claims to have spoken with monks who had spent time on 'Thule', a land to the north and west of the Faroes. There it was dark for much of the winter, though in summer there was enough light at night to allow lice to be picked off clothing. Coupled with details of the claimed voyage of St Brendan who in the sixth century sailed in search of America and saw a crystal pillar in the sea (an iceberg?) and was bombarded by burning slag from a furnace from which rivers of gold fire ran (volcanoes on Iceland or Jan Mayen?), this has led to a general belief that monks could indeed have reached Iceland by way of the Faroes. The two books also say that when the monks departed, unwilling to share their island with heathens, they left behind small bells. But neither the bells nor any evidence of pre-Norse occupation has ever been found (third century AD Roman coins have been found, but as these were traded across Europe they may well have come with the Norse) and many authorities consider the story in the two books is a rehash of Dicuil's account and may well be a myth.

In the 150 years following Ingólfur's settlement it is estimated that 1,000–2,000 people arrived in Iceland, their farmsteads occupying much of the land that was easily put to agricultural use, this chiefly on the coastal belt. By the early years of the tenth century the island was considered fully settled and to avoid conflict between themselves, the thirty-nine island chieftains instigated a parliament, the Alþing (General Assembly, pronounced Althing), at Þingvellir (pronounced Thingvellir) near Reykjavik. The Alþing sat each summer, Þingvellir therefore having a strong claim to being the world's oldest parliament site, though is not exactly the seat of democracy the Icelanders often encourage visitors to believe, the chieftains mainly discussing relationships between themselves. Justice was dispensed, and laws enacted, but these were not formally codified, being memorized by an official 'Lawspeaker'. Beneath the Alþing there were four regional councils, necessary in such a large island with a sparse and scattered population. With orally-transmitted laws and the opportunity for local bias, the system allowed for the development of injustices which in turn laid the groundwork for feuds. The Viking tradition for the rapid and violent settlement of arguments and retribution for perceived wrongs kept the Icelandic saga writers busy through the long winter nights.

Ultimately, the lack of an executive or of elected representatives at the Alþing resulted in the major landowning family fighting to achieve, and maintain, dominance. Throughout the first half of the thirteenth century Icelandic society was riven by instability. The eleventh century had also been a time of unrest as the old religion – the first Norse settlers had worshipped the Viking pantheon of gods – was replaced by Christianity. The change almost resulted in civil war, avoided only when the Alþing appointed the Lawspeaker to listen to arguments and decide on the country's future: Þorgeirr Þorkelsson, a pagan, listened, then spent a day and a night in silent contemplation under a blanket of furs. When he emerged he announced he had decided in favour of Christianity. He converted to the faith and threw his pagan idols into a waterfall: the falls, Goðafoss ('waterfall of the gods') is one of the major attractions of the northern island. The first Icelandic bishop was consecrated in 1056.

The effect of the strife over religion and the feuding of the landowners wearied the islanders, and in 1262 Iceland signed a treaty which gave Haakon IV of Norway sovereignty over the island. Iceland became part of the Kalmar Union, established in 1397 to link the three Scandinavian nations, and subsequently came under the control of Denmark when it emerged as the strongest of the three countries. Danish sovereignty did little for Iceland, offering no protection, so that an Ottoman pirate fleet was able to raid the island in 1627, taking 300 Icelanders into slavery, imposing Lutheranism against the collective will of the islanders, and imposing a trade embargo which gave Danish merchants a monopoly. The effect of the trade embargo was to impoverish the Icelanders. In the early eighteenth century smallpox killed an estimated 30 per cent of the population, then, in 1783–85, a 32km/20 mile fissure opened at Laki in southern Iceland. From it poured 12km^3 of rock, chiefly basaltic lava, a flow which covered 560km^2/220 square miles. It is the largest volcanic eruption in historical times and the largest ever witnessed by human eyes. The flow overwhelmed several settlements, but the loss of life in the eruption itself was minimal. What happened subsequently was not. Some 70 million tons of sulphuric acid were pumped out of the fissure, falling back to Earth as a rain which poisoned the soil. Several million tons of dust was also emitted: it blotted out the sun, causing crop failures in areas unaffected by the acid rain. The time is known as the 'haze of hunger', the dust causing a famine in which about 10 per cent of the Icelandic population (of about 50,000) died. At least 75 per cent of the farm animals perished. The dust spread across northern Europe and America and was responsible for poor harvests for several years: some historians believe that the poverty the dust caused in French agricultural peasants was a contributory cause of the French Revolution.

In the fifty years following the Laki eruption Iceland's climate worsened and a succession of poor harvests led to widespread hardship and significant emigration, particularly to Manitoba, Canada. At one stage Denmark considered the evacuation of the remaining population, but the idea was rejected. For those Icelanders who remained on the island, hunger, together with the iniquities of Danish rule, sparked a nationalist revival which led, in 1874, to the granting of a constitution and limited home rule. Full home rule came in 1904. In 1918 this was replaced by an Act of Union which recognized Iceland as a sovereign state with its own flag, but with Denmark responsible for its foreign affairs. The Union was due for renegotiation in 1940, but by then Denmark was under

German occupation. Iceland therefore effectively dissolved the Union, took responsibility for its own foreign affairs and declared neutrality. On 10 May 1940 British troops landed at Reykjavik in violation of Icelandic neutrality. It was an unfortunate fact of the war that the island's position, in mid-Atlantic, meant that the Allies could not take the risk of it falling into German hands. In a remarkable act, the Icelandic Prime Minister, while formally protesting the breach of neutrality, requested his people to treat the British soldiers as guests. In 1941 Britain withdrew its 25,000 troops after persuading a reluctant Icelandic parliament to accept a US occupation force – nominally an Icelandic defence force. The US set up an airbase at Keflavik and moved 40,000 men to the island: at the time the Icelandic population was about 125,000.

In 1944 Iceland held a referendum and voted to end the Union with Denmark and to become an independent Republic. In a marvellous, magnanimous gesture, King Christian X of Denmark sent a message of congratulations from still-occupied Denmark. The new government voted to renovate the island's fishing fleet, and with the income from fish sales developed an infrastructure and standard of living which is enviable.

Iceland has not joined the European Union (EU), though at the time of writing the financial crisis of 2008–9 has forced the country to reconsider this position. The original reluctance was, in part, because such a move would give all EU states rights to fish in Icelandic waters. It is a member of the European Free Trade Association (EFTA), together with Liechtenstein, Norway and Switzerland, and since 1994 has been a member of the European Economic Area (the EFTA countries except Switzerland and the EU countries). In recent years Iceland's economy has become less dependent on fishing, diversifying, chiefly, into the IT and financial sectors. Showing the traits of phlegmatism and tenacity that have allowed the islanders to overcome a thousand years and more of a harsh climate and unforgiving land, the Icelanders have achieved success out of proportion to the size of the population, which currently stands at about 280,000. Surely with such stamina they will overcome the financial disasters of 2008–9 crisis.

Icelandic agreements with Norway and Denmark over the ownership of the seas between the island and Jan Mayen and Greenland respectively are considered below.

Iceland sits on a hot plume of the Earth's mantle which has pushed the Mid-Atlantic Ridge, the spreading ridge between tectonic plates which is driving Europe and North America apart, upwards. The reason for the existence of such plumes is not well understood. They are chiefly associated with spreading ridges

(Jan Mayen, north of Iceland, sits above another) but not exclusively so: the Hawaiian islands sit above a plume well away from such a ridge. Iceland owes its existence to the plume, which has thrust land above the Atlantic's surface, but the volcanic activity associated with it has been the cause of less welcome aspects of the island's history such as the Laki eruption, the 1973 eruption on Heimaey, one of the Vestmannaeyjar (Westmann Islands) which lie off Iceland's southern coast, which formed a new mountain and required the evacuation of the inhabitants, and the 1996 eruption of Grímsvötn beneath the Vatnajökull ice cap which created a *jökulhlaup* (glacier flood) which released an estimated 4km^3 of water and 1 million tons of ice onto Iceland's southern coast: again the eruption was foreseen and no lives were lost.

For the volcanologist, Iceland is an excellent study centre, with excellent examples of shield, strato- and subglacial volcanoes, spatter rings and cinder cones, geysers, lava tubes, pseudocraters and more. Life scientists have been able to observe the way in which plant and animal life has developed on Surtsey, the Earth's newest land, formed by a sub-sea eruption off the Westmann Islands from 1963–6. For the visitor, Iceland's position and volcanism offers dramatic and breathtaking scenery. Though sub-Arctic – the Arctic Circle cuts through the island of Grimsey off the northern coast – volcanic land-building has raised the island's height sufficiently for glaciers to have formed, and the climate is such that there is limited flow to sea level: at Jökullsarlon, in southern Iceland, a glacier calves small icebergs into a tidal lagoon. Volcanic activity has created tall, beautiful waterfalls, areas where mud pools bubble and the air is filled with the stench of bad eggs from venting hydrogen sulphide, extensive lava fields covered in mosses thick enough to hide a walker's boots, and other joys. At Þingvellir, not only can the site of the Alþing be viewed (one of the three sites of the near-compulsory 'Golden Triangle' offered by just about every island tour company: the other two being 'Strokkur', the remaining active geyser at Geysir, the site for which all the world's geysers are named – if not spelt – and the superb Gullfoss waterfall) but the visitor can walk a path through the spreading gap of the Mid-Atlantic Ridge. As the ridge separates the North American and Eurasian tectonic plates, the plates are one on each side, an extraordinary experience.

Set in mid-ocean, and with peaks high enough to influence airstreams, Iceland's climate is highly variable, so the visitor may peer through rain and mist at the curious landscape, or may see it with amazing clarity. Free of large industrial complexes, and with a limited population, the atmosphere of Iceland is largely free of the pollutant haze which is often a feature of other European countries. The island is also dotted with National Parks which preserve stunning vegetation and the nesting areas of Arctic and sub-Arctic birds. The combination makes the island an essential visit for lovers of wilderness and birdlife of the Arctic fringe.

But the natural wonders of Iceland which attract the visitor also offer the potential for exploitation. The Icelanders have always exploited the geothermal energy beneath the island, harnessing venting hot water for heating and electricity production. As the production of greenhouse gases by the burning of fossil fuels has now been recognized by mainstream science as a significant contributor to global climate change, Iceland's use of alternative energy is a major bonus for the island (which has, in any case, no potential for mining its own fossil fuels). With a rising population and a desire to continue the move away from an economy based on fishing (an industry which is facing its own problems), the Icelandic government has sought ways of utilizing both geothermal energy and the island's landscape. Greater use could certainly be made of geothermal energy sources, but the concentration has been on the damming of valleys for hydro-electric schemes. Some long-term suggestions involve the laying of sub-sea transmission lines to connect Iceland with Britain, using DC rather than AC to limit line losses. More immediate plans invite electricity intensive industries to Iceland to use the power at source.

A beneficiary of the latter course are aluminium producers, the metal being produced by electrolysis rather than by conventional smelting. The biggest project to date is the Kárahnjúkar power station to the east of the Vatnajökull glacial ice cap, which, when completed, will harness the flow of two glacial rivers and feed it 50km/30 miles by two 220kV transmission lines to power an Alcoa aluminium smelter at the head of Reyðarfjörður. Alcoa is the abbreviated name of the Aluminum Corporation of America, a multi-national company which operates in fourty-four countries and is the third largest producer of aluminium. The Reyðarfjörður scheme is the largest construction project ever undertaken in Iceland. At first glance the use of greenhouse gas free power is a positive, particularly as no one will be displaced by the project, and the world needs aluminium. But the project has a considerable downside. The original planning application for the project was rejected on environmental grounds, but this decision was overturned on appeal, the government insisting that it made sound economic sense to bring employment to an area that was seeing gradual depopulation, as well as utilizing renewable, 'green' energy to produce a metal essential to modern life.

Objectors point out that the 1,000km²/390 square miles of wilderness (about 1 per cent of Iceland's land area) which will be directly affected by the project –

being drowned by the reservoir behind the five massive dams under construction, disappearing under roads or disturbed by related works or by stream diversion – is one of the most important and sensitive in the country. Calculations suggest that up to 3 per cent of the country's land area might be affected by wind-blown soil from the site. This can dry wetland areas, and, as the vegetative ecosystem over much of the country is fragile, that could have a significant adverse effect. The affected area is an important bird breeding ground for Arctic and sub-Arctic species, some rare in Iceland. The sediments that were once washed out to sea previously formed a bar which hindered coastal erosion and which formed a moulting ground for geese and a breeding ground for seals. Ironically, the now trapped sediment will, eventually, block the water intake for the power station, shutting it after a period of 50–80 years. Geologists also worry about the sheer weight of water behind the massive dams and how it will affect a country renowned for its instability. Another problem is that the dams are located on an area of faults in the Earth's crust, making the project costly, as filling cracks in the seismic active zone is expensive.

There are other factors too which make even those reconciled to adverse environmental effects concerned. Alcoa's licence from the Icelandic government allows it to emit 12kg/26½lb of sulphur dioxide per manufactured ton of aluminium, far in excess of the World Bank 'limit' for such plants. The licence therefore means that Alcoa are using not only Icelandic electricity, but a large portion of the 10 per cent increase in permitted emissions that Iceland negotiated under the Kyoto Protocol, an allowance granted because of the island's 'green' power, limited population and limited industrialization. Aluminium production also releases tetrafluoromethane (CF_4, also called carbon tetrafluoride) and hexafluoroethane (C_2F_6) – which are not only extremely potent greenhouse gases but, crucially, extremely stable, with atmospheric lifetimes of about 50,000 and 10,000 years, respectively. Bauxite, the raw material of aluminium production, will also need to be brought to Iceland by ship, which has an environmental cost. Of more concern is the fact that an increase in primary aluminium production is also difficult to justify when recycling old drinks cans is so much cheaper and there is no world shortage of the metal. The green argument for production in Iceland is also called into question when Alcoa's decision to close other smelters in the United States is recalled: the company is moving to Iceland, it seems, because it is cheaper to manufacture there and the pollution control is less stringent. For all the green colour splashed on the Icelandic decision, Alcoa's green credentials are not wonderful. Its Rockdale, Texas plant burns strip-mined lignite to produce electricity and its history as a major polluter is far from enviable. The company's involvement in the Barra Grande Dam on the Pelotas River in Brazil has also been controversial, with claims of the production of fraudulent environmental impact reports.

Closer to home, the Icelandic government's own case on the creation of jobs has also been questioned. There is limited unemployment in Iceland and there are likely to be few takers for dirty, risky jobs which are at odds with the way Icelanders wish to work. The majority of construction work has been carried out by imported labour, chiefly from eastern Europe (official figures suggest 80 per cent are Polish) and objectors feel the same will be true of jobs at the smelter plant as well.

Iceland's problem encapsulates a world problem, one which is now advancing towards the Arctic. Despite all the warnings on climate change, the world's industrial giants pursue the strategies they understand and which have worked consistently, supplying shareholder dividends and management bonuses. They will always seek ways of maximizing profits, taking advantage of any reduction in costs and exploiting any weakness in local legislation. The business known as 'Iceland' is pursuing an essentially similar strategy, even if in this case the shareholders are the general population which will undoubtedly benefit from the wealth generated by the use of the country's green power. But the fear must be that the inherent advantages of Iceland's strategy will be lost in the rush to seize any opportunity to make money, turning a wonderful environment into yet another industrial backyard.

Jan Mayen (Norway)

In 1607, attempts to discover either North-East or North-West Passages having failed, English sea captain Henry Hudson attempted to reach the Orient by sailing due north, over the Pole. Today the idea seems ludicrous, but at the time the scientific view was that the perpetual sunlight of the Arctic summer would melt the ice at the Pole, leaving a ring of ice, where the period of the midnight sun was too short for complete melting, surrounding an open polar sea. All Hudson needed to do was search for a breach in the ice ring and sail through, crossing the open sea and searching for another break in the ice on the eastern side. In the *Hopewell*, with a crew of just eleven, Hudson followed the east coast of Greenland before striking out north-west for the newly discovered Svalbard. There he turned north again, eventually reaching 80°23′N, a northing that would not be bettered for more than 150 years. But the ice ring was continuous and Hudson turned the *Hopewell* around. On the voyage south the wind pushed the ship west and Hudson sighted a volcanic island which lay to the north of Iceland: he called it Hudson's Touches.

It is likely that Hudson's chosen name was unknown to the Dutchman Jan Jacobsz May van Schellinkhout, who is the first person known to have landed on the island when he captained one of two whaling ships which arrived in 1614. For the next thirty-five years the Dutch hunted whales around the island (see page 120), but then the island was largely ignored for two hundred years until, in 1882–3, it was the base for the Austro-Hungarian expedition which was part of the first International Polar Year. There were further scientific visits in the years that followed, but more sustained occupation waited until 1906 when Norwegian trappers arrived, word having reached them of the island's substantial fox population, the majority of the animals being 'blue', meaning their pelts could be sold at a premium. In 1921 the Norwegian government set up a meteorological station on the island. Then, in 1929, Norway formally took possession of the island. The weather station was manned permanently until late 1940 when the staff were evacuated and the station destroyed by the British who feared it would be taken over by the Germans. Norwegians returned in the spring of 1941, Jan Mayen becoming 'Free Norway' (Norway had been occupied in April 1940). From 1943–5 an American radio base occupied a site close the Norwegian base. In 1959 a LORAN station was set up on the island to aid air traffic across the Atlantic. Both the weather station and air traffic radar stations remain – though they are now on the south side of the island. They are manned by Norwegian personnel who spend six or twelve months in residence. The stations are serviced by aircraft. Visitors are not permitted on these (infrequent) flights, and commercial aircraft are not able to use the airstrip. The only way for visitors to reach Jan Mayen is therefore by sea, and there has been an increase in Arctic tour vessels visiting the island in recent years.

As with Iceland, Jan Mayen is situated over a plume hot spot of the Mid-Atlantic Ridge, but here plume activity has created a much smaller island. Jan Mayen is spoon-shaped, 54km/33½ miles along its long axis, and only 2.5km/1½ miles wide at its narrowest point. It is dominated by Beerenberg (Bear Mountain), a 2,277m/7,434ft volcano from whose flanks glaciers descend into the sea. The visitor is rarely treated to a glimpse of the volcano's summit as it is usually shrouded in cloud. What are visible are lava fields, usually black – here, as in places on Iceland, the sand on the beach is also black – but dotted with patches of red iron oxide, and jagged lava cliffs. And again as on Iceland, the lava fields are home to mosses, lichens and liverworts. In places the plant coverage is almost total, in others the emerald green splashes of colour on the black lava seem almost surreal: Jan Mayen is a showcase for the tenacity of life.

With the fox population now long extinct, Arctic bird species are the only wildlife on show. Ship-borne visitors are unlikely to have much time to explore the seabird cliffs, but are usually able to spot whales in the waters close to the island.

In 1981 the governments of Iceland and Norway signed an Agreement on the delineation of the maritime border between Iceland and Jan Mayen in the wake of the Icelandic decision to extend its territorial waters to 200 nautical miles (320km). The agreement specifically addresses the box formed by 70°35′N, 68°00′N, 10°30′W and 6°30′W where the two islands lie within 400 nautical miles (640km) of each other. Within that area, each country has been granted a 25 per cent stake in mineral resources (the Agreement specifies petroleum by name) discovered in the other country's section of the box.

Fennoscandia

As we have seen, first contact between the Sámi and the early Norse settlers of southern Fennoscandia probably occurred in about AD 500. For five hundred years the contact was probably spasmodic and trade based, though it is clear that the Norse were gradually moving north, particularly along the western coast of Norway, then as now the North Atlantic Drift pushing the Arctic further north than in the Nearctic. It is worth noting here this very great difference between the position of the 'true' Arctic and, say, the Arctic circle on the two sides of the Earth. In northern Norway there are vibrant cities, supported by agriculture and industry, as far north as 71°N, while paved roads aid visitors to reach Nordkapp, the notional northern point of Europe. In North America comparable towns are not found north of the Arctic Circle (though there are some truly vibrant settlements of native northern dwellers), while at 71°N paved roads are a very distant memory.

As well as substantial cities, northern Fennoscandia has an abundance of beautiful scenery. Norway has a central mountain chain, the north being a lower-lying land deeply cut in places by the fjords that have made the country famous. By contrast, northern Sweden is more mountainous. Finland and the Kola Peninsula are low-lying, vast areas of forest being interspersed with lakes. All four countries have created National Parks to protect the best of the scenery and wildlife areas.

As a consequence of the northern expansion of the Norse and, in western Russia, of the Novgordians towards Karelia, the Sámi found themselves under increasing pressure. They were now being taxed, in some cases by both western and eastern neighbours, and their traditional way of life was threatened by the advance of Christianity. When there was conflict between the Norse and the

Novgordians, Sámi settlements were occasionally drawn into the conflict as they stood on disputed territory.

Ultimately, the emerging Scandinavian kingdoms and, after the rise of Muscovy, Russia, divided the territory of the Sámi between them. While still in principle maintaining their traditions, the Sámi were now part of kingdoms not of their choosing. Their position within the kingdoms was also that of subjects, second-class citizens who were seen as little better than savages and could be treated as such. When Sweden banned slavery in the mid-fourteenth century the legislation did not cover the Sámi as heathens' personal liberties and freedoms did not apply. There is even evidence to suggest that the efforts of missionaries were reined in to avoid the complication of dealing with Christian Sámi. Taxation could also be levied more indiscriminately against a pagan community, and while most Sámi were taxed by the kingdoms in which their communities stood, the northern Sámi were often taxed by Norway, Sweden and Russia, or (more often) by two of the three. Later, during the expansion of the Swedish empire in the late-sixteenth/early-seventeenth centuries, Sámi settlements in northern Finland and on the Kola Peninsula were attacked and destroyed by both sides, apparently for no reason other than being in the way when armies passed. At that time, Sweden's primary aim was control of the Baltic and its trade routes, its armies taking Finland and the Baltic states, and forcing the Danes to give up southern Sweden. In the early seventeenth century a border between Sweden (which then included Norway) and Russia was agreed in Finnmark so as to exclude each other's tax gatherers.

Yet apart from tax gathering and the occasional, but obviously unpleasant and unwarranted, committing of atrocities on them, the Sámi continued with a lifestyle which was little changed for centuries. They lived in *siida*, communities, which each held a defined territory and was controlled by a chief. The chief or, occasionally, the chief in conjunction with a community council, allocated hunting, trapping and fishing areas to individual families. The community also acted in unison in activities that required increased labour, such as the fishing of spawning salmon, hunting in winter and, on the coast, the hunting of migrating whales. Animals were hunted and trapped for food and skins, the latter being used for clothing and also as trade goods. Taxes were also paid in pelts. The nature of the tax is interesting because it is not immediately obvious why the Sámi would consent to taxation by a foreign king. It seems that initially the tax was a 'gift' paid in order to be allowed to continue to live unmolested in an area now having been sequestered by a formidable southern king. Later the tax was levied as much to protect the Sámi from being pestered by the tax collectors of another king. No matter how this system is portrayed, 'protection money' would seem to be the most straightforward definition. Within the *siida*, there had always been an element, at least, of families making contributions to the community, so the imposition of taxes which were removed from the *siida* led to a reduction of its authority and autonomy. The rise in reindeer husbandry in the seventeenth century, which allowed families to become more independent of the community, also mitigated against the power of the *siida*, though in some areas the *siida* lasted until the early twentieth century.

Herding was not taken up across the Sámi homeland with equal vigour, the practice starting mainly in the west and spreading only slowly eastwards: herding was not common on the Kola Peninsula until the nineteenth century, and only then because of an immigration of western Sámi who brought the tradition with them. This is a fascinating reversal of what would be expected, as there is evidence to support the view that reindeer herding developed in the eastern Arctic and spread west. Today, only about 10 per cent of Sámi herd reindeer, despite the perceived view that it is a commonplace.

The rise in reindeer husbandry coincided with a more concerted attempt at converting the Sámi, the earlier lack of enthusiasm for the allowing missionaries to head north being replaced with a zeal that belied the fundamentals of the faith the missionaries were espousing. As with all the Arctic native dwellers, the Sámi adhered to an animalistic religion, each *siida* having a *noaidi* (shaman). The attitude of the Finnish missionary Gabriel Tuderus (1638–1705) to the Sámi was common, if not always held so vehemently: he described them as 'flesh-coveting swine and curs incited by Satan to resist the true, pure and solemn sacraments'. It is no surprise that violence towards shamans and the violent conversion of individuals was not uncommon. The spread of Lutheranism throughout Scandinavia, and the scientific 'Enlightenment' of northern Europe aided a gentler absorption of Christianity.

In the eighteenth century the borders between Norway (then part of the kingdom of Denmark), Sweden and Russia (the latter including Finland which was ceded by Sweden to Russia in 1809: the Sámi population was reduced by starvation and disease as the remnants of the Swedish army retreated) were drawn up. That between Norway and Sweden included a codicil which gave the Sámi the right to trade, herd, hunt and trap across the northern borders as they had always, but disallowed the right to own land on each side of the border. Migration of the reindeer herds meant that there was considerable traffic of Swedish Sámi into (and out of) Norway, lesser traffic of Norwegian Sámi into Finland, and lesser again of Finnish Sámi into Norway. There was virtually no traffic to and from Russia's Kola Peninsula. One often overlooked aspect of the reindeer herding lifestyle is that it was not taken up by all Sámi. On the coast, and to a lesser extent inland, the traditional life based on fishing, hunting and trapping continued, but the wealth (and therefore status) of the herders, together with the romantic view of the semi-nomadic herding life, meant that the non-herding Sámi became marginalized, a downmarket and not-entirely-authentic Sámi, a status from which non-herders (and there are still many of them) have never really escaped, a fact which causes friction between those following the two lifestyles.

In the later years of the eighteenth century the Sámi came under pressure from an influx of southern settlers, some of whom used the courts to uphold the acquisition of traditional Sámi land. Some settlers also produced grain alcohol which they traded with the Sámi, with predictable results. The tensions between the Sámi and the incomers occasionally led to conflict: in 1852 the closure of the Norwegian-Finnish border was a final straw for the Sámi of Kautokeino (still the most significant Sámi village in Norway, indeed in Fennoscandia) and in a riot a Norwegian law officer and a Norwegian liquour merchant were killed. The 'wild men', as they were called, who had sparked the riot were arrested: two were executed for murder, fourteen received life sentences and the forfeit of all their possessions. The son of one of the instigators was beaten to death and a daughter died in prison. In a particularly insensitive act, the authorities chose to behead one of the two condemned men and gave the head to a university for scientific research.

The final drawing of international borders also affected Sámi society, the Russians closing Finland's borders, cutting off migration routes: only the Norwegian-Swedish border remained essentially transparent. On the Kola Peninsula the rise of Russian monasticism was detrimental to the Sámi, the monasteries taxing them heavily. When the monasteries were dissolved, in part 'because of the monks' evil ways', the situation did not improve because rich merchants moved into the area, dominating trade, which of necessity was mostly sea-borne, to the further detriment of the Sámi.

All the states of Fennoscandia also instigated a programme of assimilation of the Sámi which included the teaching of the state language. Although the official reason for this was that to take full advantage of education and welfare programmes it was necessary for the Sámi to learn the majority language, there is little doubt that fear of 'foreigners' in the sensitive border areas of the north were at the heart of the policy: in Norway's border areas, land sales were also restricted to Norwegians. Although in Norway and Sweden, use of the Sámi language was not actively discouraged, at least not officially, on the Kola Peninsula, use of Russian rather than the Sámi language in schools was almost universal. However, this was less for ideological reasons – though doubtless there was an element of that – than for the fact that there were so few Sámi speakers qualified to teach it. However, it was in Russia in 1935 that the first attempt was made to produce a written Sámi alphabet and grammar. The teacher behind the attempt died and progress was halted until 1982 when the project was finally completed and teaching in Sámi began. By contrast, the three Scandinavian countries did not recognize Sámi as an official language until the 1990s.

In the early years of the twentieth century there was a change of attitude in Sweden and, to a lesser extent, the other Fennoscandia nations, the idea being that assimilation had not been successful, so the Sámi should be allowed to get on with their reindeer herding while the rest of the nation progressed towards a modern future, an attitude which has also been seen in the Nearctic, one which turns the native Arctic dwellers into a theme park attraction. The 1939–45 war interrupted any further progress, though the redrawing of the Soviet-Finnish border in 1944 caused many Sámi to move west from Suenjel/Petsamo as they had no wish to stay in the Soviet Union. After the Finns and Russians had fought the 'Winter War' of November 1939–March 1940 Finland and Germany concluded an agreement which saw 220,000 German soldiers in Finland, many in Sámi lands, and both Finnish and German troops attacking northern Russia. When a truce was agreed between Finland and the Soviet Union in 1944, Finland lost territory, including its corridor to the Barents Sea. It also then fought to expel the Germans from Sámi lands, the Germans retreating to Norway, but using a 'scorched earth' policy to cover its retreat, destroying much of the infrastructure of Sámi Finland.

After the 1939–45 war an acceptance of the value of Sámi culture, history and language grew within the three Nordic countries, though the idea of assimilation guided government policy through to the 1970s. A particular issue was land and water rights, something which was was brought into sharp focus by the dispute over the Alta-Kautokeino river scheme in the 1970s. This hydro-electric scheme required the building of a massive dam, the reservoir behind which would drown a great deal of prime reindeer herding land. The Norwegian government view was that the benefits of the scheme to Norway were clear and obvious, and was genuinely shocked by Sámi outrage. A demonstration by Sámi protesters was forcibly broken up by police in what seemed to many Norwegians a very heavy-handed approach. Then seven Sámi went on hunger strike in a *lavvu* pitched illegally on the lawn outside the parliament building, causing the government considerable embarrassment, while Niillas Somby, another Sámi activist, attempted to blow up a bridge near the site of the proposed dam. Somby lost his hand in the explosion and a photograph of the severed hand, placed on the Code of Norwegian Law, became a powerful symbol of the Sámi cause. The depth of feeling forced the government to a realization that this was not just a group of curious northern folk who seemed intent on resisting all attempts to move into the twentieth century, but a distinct people with a valid culture, and a valid interest in the development of their homeland. The river scheme went ahead, but one outcome of the confrontation (though it took nine years to reach fruition) was the creation of the *Sámediggi*, a parliament with control over many aspects of the governance of Sámi land. A similar parliament was formed in 1993 in Sweden, a surprisingly late date in view of the greater grievances the Sámi had with the Swedish government, particularly over the loss of land for hydro-electric schemes and the position on the compensation for such losses (which went to a State-controlled 'Lapp Foundation' rather than the individuals concerned, the use of 'Lapp' in the name being an additional insult). Even later was the equivalent parliament in Finland, which did not come into being until 1996. Each of these parliaments are backed by a number of organizations formed to further the aim of ensuring the flourishing of Sámi culture. One of these is *Sámiid Duodji*, which protects and develops traditional handicrafts: visitors to Sámi areas should look for the *duodji* sticker to ensure they are buying genuine crafts.

The position of the Sámi of Russia's Kola Peninsula is poorer than that of their cousins in the Nordic countries. At the time of the Russian revolution it is estimated that about 9,500 people lived on the Peninsula, though probably less than half these were Sámi, the remainder chiefly Komi, an ethnic group from the north-west Urals who had moved to Kola in the nineteenth century and taken up reindeer herding. During the first half of the twentieth century the population had risen to almost 250,000. By the 1990s it was more than 1,200,000: it is now just below one million. The effect of this influx, and the industrialization which produced it, was to make the Sámi a very small minority in a land which had changed immeasurably, often for the worse: the pollution plume from the Nikel smelters has created a wasteland downwind. Today there are only about 2,000 Sámi on the Peninsula (in comparison to about 40,000 in Norway, 20,000 in Sweden, and 7,500 in Finland). Although one of Lenin's aims was the promotion of the culture and language of Russia's minorities, Stalin had a very different view, suppressing the use of the Latin alphabet and imposing the Cyrillic, and teaching in Russia (though there was an attempt to create a Sámi alphabet based on Latin script in 1933, an attempt abandoned in 1937). Today less than a third of Russian Sámi speak their mother tongue.

With the forced collectivization of reindeer during Russia's communist era, sales were allowed only through the *kokhoz*, the state office, which also sold life's essentials, so the herders were effectively tied to the system. This was a further hardship for the Kola Sámi. With communism's collapse the situation has, for many, worsened. In an irony that would be comical if not so deadly serious, the Sámi were offered the chance to buy their herds. With limited capital at their disposal, few available market outlets for the meat, and land rights open to dispute, those inclined to do so found it hard to take up the offer. Many ordinary Russians also found the concept of privatization by voucher difficult to understand and so too readily sold their vouchers, allowing ownership to be taken up by former system managers, who did understand the system and acquired businesses for themselves. Private entrepreneurs with money also moved onto the Peninsula, sometimes buying up the rights for salmon fishing on traditional Sámi rivers: the Ponoy River, the Kola's primary salmon river, is now leased to a private outfitter, the Sámi being forbidden to fish a river that historically was theirs. A Sámi Association was formed in 1989 and other organizations have also been set up, but there is nothing comparable to the Nordic *Sámediggi.* The Russian Sámi are also members of RAIPON (see pages 65–6).

Cross-border Sámi organizations have existed since 1945, the most important of which was the Sámi Council. This began life as the Nordic Sámi Council in 1956, but in 1992 was extended to include Russian representatives. Then in 2000 a Sámi Parliamentary Council, with representatives from the three Nordic parliaments, was formed. The move has been met with concern by the three central governments because as pressure for forestry, hydro-electric production and mining in Sámi areas increases, the Council's references to a united Sámi nation (known as Sápmi) has worrying connotations.

Svalbard

An Icelandic manuscript of 1194 includes the notice *Svalbarði fundinn* – Svalbard found. Exactly what had been seen by the Norse on the voyage that gave rise to the proclamation is still debated, and has rarely been more important than it is today. *Svalbarð* clearly derives from *Sval barð* – cold edge – but which cold edge? Iceland's *Landnámabók* (see above) adds further detail – *frá Langanesi á norðanverðu Islandi er iiii. Dægra haf Svalbarða norðr I Hadsbotn* – from Langanes on Iceland's north coast it is four days sailing to Svalbard at the northern edge of the ocean. Norwegian historians tend to the view that that the brilliant seamanship of the Norse, together with their known northern voyages in search of walruses, whose skins provided the strongest ropes, mean that they may well have chanced on the archipelago now known as Svalbard. Russian (and other) historians take a different view, suggesting that the sailing time and direction to the Norse Svalbard means it was much more likely to have been Jan Mayen or the eastern coast of Greenland. The Norwegians fear, correctly, that the Russian view of what was meant by 'cold edge' is related to their own ownership claims for the archipelago; Russian historians claim that prior to the official discovery by the Dutch in 1596, Pomors had already visited the islands. There is good evidence to support Pomor visits in the early eighteenth century, as traditional Russian crosses have been found in several places in the eastern archipelago, together with the remains of timber huts and graves. Dendrochronology of the wood suggests dates as early as the mid-sixteenth century, but as logs carried downstream by Siberian rivers are frequently washed up on Svalbard beaches, that is far from conclusive. However, there is a tradition in north-western Russia that the Starostin family of Pomors were hunting on Svalbard before the construction of the Solovetsk monastery (on an island in the White Sea) in 1425. There are also documents that note that the Russian Tsar had taken possession of *Grumant* in the mid-sixteenth century. The name translates as Greenland, but as Svalbard was initially considered to be part of Greenland that could explain the confusion. There seems little doubt that the Pomors had the technology to reach Svalbard. Their ship was the 'koch': small, with curved sides and a flat bottom. The ship had a single mast, but was light enough to be rowed by the small crew. This, and the shallow draft, made them highly manoeuvrable in icy waters, the flat-bottom allowing then to be readily freed if pack ice threatened entrapment. Kochs had a double hull, an inner layer of boards sewn together with juniper roots, and a nailed outer layer that gave them excellent resilience. Semen Dezhnev used kochs on his journey through the Bering Straits in 1648, each of the seven ships having a crew of between ten and fourteen. Pomor visitors to Svalbard also had the skills necessary to deal with the hostile environment, being able to live for weeks on their ships and almost indefinitely off the land. In 1743 four men were accidentally marooned on Edgeøya, the island off Spitsbergen's eastern coast, and survived for six years by hunting, drinking blood to ward off scurvy, the one man who objected to drinking blood dying of the disease. The survivors were said to have been in good health when they were rescued. While this event was several hundred years after the suggested first Pomor journeys to Svalbard, technology had advanced little in that period.

It is likely that no conclusive evidence for Norse or Pomor journeys to Svalbard will now be discovered, allowing the first unambiguous journey to be that of the Dutch expedition of 1596, which included Willem Barents. Barents reached and named Spitsbergen, the archipelago's largest island (named for the 'pointed peaks' the Dutch noticed), and actually sailed to its north-west tip at 79°49′N, almost certainly a record northing at that time. During the journey

Barents' expedition also reached Bjørnøya (Bear Island), the most southerly island of the archipelago, naming it after the polar bear which they killed close by.

Following the discovery of the abundance of whales in Svalbard's waters, the Dutch, British, Basques, Danes and Norwegians all visited the archipelago on whaling expeditions, with the Dutch building the largest and most famous whaling station, Smeerenburg (Blubber Town) on Amsterdamøya off Spitsbergen's north-west coast. The slaughter of the whales led to an abandonment of Svalbard in the early eighteenth century though groups of Norwegians and Pomors still visited to hunt and trap. At first the Pomors were in the majority, but over time they lost interest in the archipelago, leaving it to the Norwegians who represented a more or less continuous presence.

Scientists arrived in the nineteenth century. The Swedes were particularly active, and the expeditions of Adolf Erik Nordenskiöld are famous. The remains of his observatory can still be seen at Kapp Thordsen, Isfjorden, Spitsbergen. In the late nineteenth century Svalbard's proximity to the North Pole encouraged expeditions to use it as a starting point. Of these expeditions the most famous is the tragic Andrée balloon expedition; the remains of the three balloonists were found on Kvitøya, at the north-eastern edge of the archipelago, thirty-one years after they had disappeared. Film in the men's camera, preserved by the cold, could still be developed, revealing some details of their struggle to survive after the balloon had come to grief. Remnants of the men's last camp can still be seen, though Kvitøya is not readily accessible. The remains of later airship attempts to reach the Pole are more easily reached: those of the American Walter Wellman at Virgohavn, and the mooring mast of the airship *Norge*, in which Roald Amundsen and others overflew the Pole in 1926. The mast was later used by the airship *Italia* in a 1928 flight, which ended with the deaths of some of the crew and a rescue from the ice: in one rescue bid, Amundsen lost his life.

In 1896 Sir William Martin Conway made a crossing of Spitsbergen, opening the way for many later adventurers. Two years later, to cater for the number of visitors wishing to find similar excitement, or merely to experience the Arctic, a basic hotel was built beside Adventfjorden. The following year coal was mined commercially for the first time. Then, in 1901, the American John Munroe Longyear came as a tourist. Realizing the potential of the area, Longyear bought an existing mining company and began the town which now bears his name – Longyear's Town or Longyearbyen, at 78°N. Before Longyear's arrival coal mining had been exclusively Norwegian, but after his arrival the British, Germans and Russians began to take an interest. Concerned that disputes over mining rights were creating chaos on its northern doorstep, in 1907 Norway declared a claim on the archipelago. The basis of the claim was interesting as it was based less on Norse discovery than on the fact that Norse occupation of Greenland was adequate as it had once been assumed that Greenland and Svalbard were joined. In addition, in the seventeenth century the Danish king (at a time when Danish sovereignty included Norway) had attempted, unsuccessfully, to tax Dutch and British whalers on Svalbard. Once whaling had ceased, the Danish/Norwegian claim had lapsed as the archipelago seemed devoid of worthwhile resources. A conference was organized, with delegates from America, Belgium, Denmark, France, Germany, Great Britain, Holland, Russia and Sweden, but such was the time spent organizing that the 1914–18 war had broken out before the conference convened. When peace returned, the major European powers were reluctant to engage in any discussion which might result in one of them losing control. It was therefore deemed appropriate that the Norwegian claim to sovereignty be ratified. This was agreed in 1920 and the Treaty of Svalbard was signed in Paris. In 1925 Svalbard formally became part of the Kingdom of Norway and the Treaty came into force. The Treaty is reprinted in Appendix 1. The most interesting aspects of the Treaty are the granting of equal rights in terms of 'maritime, industrial, mining and commercial operations' (Article 3) and 'right of ownership of property, including mineral rights' (Article 7). Article 8 stated that Norway would provide mining regulations for the archipelago. This they did, producing the Mining Code (Mining Regulations) for Spitsbergen (Svalbard) in 1925, amended in 1975.

Article 9 of the Treaty required that no military activity should be carried out on Svalbard. The Article was ignored during the 1939–45 war when Spitsbergen became an important Allied meteorological base, transmitting weather reports for the benefit of the Arctic convoys en route to Murmansk. Mining operations had been suspended and the mines destroyed to prevent them falling into German hands. These activities, part of Operation Gauntlet in 1941, involved British, Canadian and Norwegian forces. Weather stations were set up at both Longyearbyen and Barentsburg. Each was attacked by German forces, from the air in May 1942 and from the sea in September 1943. The latter attack was by a fleet of nine destroyers and the battleships *Tirpitz* and *Scharnhorst*. The attack, codenamed Zitronella (or, apparently, Sizilien), was aimed at taking over the weather stations. Ground troops were landed, but it was very soon apparent that they could never be adequately supported and they were withdrawn. However, the Germans did set up a weather station on Hopen, an island to the south-east of the main archipelago which remained operational until 1945. Since the end of the war Norwegian military ships and aircraft have visited Svalbard, but no permanent base has ever been set up.

With the cessation of hostilities, both Norwegian and Russian coal mining activities resumed, the provisions of the Svalbard Treaty being overseen by a Norwegian *Sysselmannen* (Governor) at Longyearbyen, with mining issues overseen by the *Bergmesteren*. The system has worked well, though there was a dispute following Norway's decision in 1976 to establish a 200-nautical-mile exclusion zone around its coast, and to create the Spitsbergen Fishery Protection Zone, a 200-nautical-mile zone around Svalbard within which Norwegian fishermen would have sole access. The creation of the zone was disputed by other signatories of the Treaty on the grounds that Articles 1 and 3 gave all signatories equal rights of both fishing and maritime activity. Norway argues that these rights did not envisage a free-for-all that endangered fish stocks and points to Article 2 which gives Norway the freedom to instigate measures which ensure the preservation of fauna within its territorial waters. Having formally implemented the sovereignty which the Treaty affords it (something which Norway did not actually do until the issue of the Zone arose), Norway considers its obligations under the Treaty take precedence over rights to others. At present, in general, the fishing fleets of other nations acknowledge the Zone.

A much more important issue has arisen from the establishment of the Zone: the international border between Norway and Russia. The fixing of this is crucial to the exploitation of the Barents Sea (by either nation). Norway maintains that the 4-nautical-mile (7km) limit on territorial waters inherent in the Svalbard Treaty are irrelevant in the fixing of Norway's 200-nautical-mile maritime territorial limit. Russia disagrees. A further contentious issue is the Russian opinion that its 'regime' in the Arctic is defined by the 'sector principle', giving it jurisdiction over all the area north of its mainland coast within a triangle whose three points are the North Pole and the eastern and western limits of the Russian mainland. The difference of opinion creates a large 'grey area' of Barents Sea.

On Svalbard, Longyearbyen, the archipelago's 'capital', has increased dramatically in size in recent times, with the building of a University Centre (the world's most northerly educational institute), which offers courses in Arctic biology, geology and geophysics as well as research programmes, and the building of hotels and other tourist facilities. There are many opportunities to join ships offering tours of the best sites on the archipelago. In winter, where once the opportunities were limited to a few companies offering snow scooter trips, the independent traveller can now hire a scooter from one of several companies. So popular has travel to Svalbard become that restrictions on travel have had to be introduced, and visitors have had to be policed to prevent the kind of behaviour that endangers both the wildlife and the visitors themselves: polar bears are fully protected on Svalbard which makes them even more fearless than, as top predator, they are by nature, and this has led to several very unfortunate incidents.

The Svalbard archipelago comprises Spitsbergen and several nearby islands, including the larger islands of Edgeøya and Nordaustlandet, and smaller islands near each of these; the island group comprising Kong Karls Land; Hopen; and Bjørnøya – Bear Island – to the south of Spitsbergen, which lies roughly midway between Svalbard and the Norwegian mainland. Though remote from the main archipelago, Bjørnøya is geologically similar to west-central Spitsbergen and is included in the Svalbard Treaty. Bjørnøya is a low-lying island and one that seems to attract cloud cover, a combination which meant that before the era of GPS systems it was frequently missed by those seeking it. Many other islands of the archipelago are similarly low-lying, but Spitsbergen, by far the largest island, has a very mountainous central area.

Russia

In comparisons to the other Arctic nations, Russia has had a very complicated history. It has suffered slavery under Mongols, developed a sovereign monarchy, and been infected with communist ideas from which it emerged to find itself in the hands of oligarchs and secret factions. Throughout this history, Russia has interacted with the Arctic, sometimes investing in it, at others neglecting it, but seemingly always submerged in an array of Arctic problems.

Between the fourteenth century to the first half of the sixteenth century, the Moscow state rose from a vassal Mongol state – a conglomerate of separate feudal societies – to one dominating what may now be termed European Russia. Novgorod and, to a lesser extent, Kiev, were the only large territories with maintained trade routes, the rest of what is now Russia being agricultural. Most of population in the countryside were serfs (essentially slaves) owned by a landowner. They did not pay taxes, working exclusively for the landowner who could sell his/her serfs (souls, as they were called) if he/she wished. The only people who enjoyed any form of freedom were the cossacks. Today the word has a romantic air, but it derives from the Turkish *kazak* and in sixteenth century Russia denoted those who lived by their wits at the edge of the realm and at the boundary between law and order, civilized life and banditry. Cossacks worked in agriculture as free men in exchange for having to protect the borders from enemy incursion, a position which altered when Sweden attacked Russia, requiring Peter the Great to establish a regular army.

Agriculture work is a highly seasonal business in Russia. Summer was a busy time, but in winter it was customary for men to leave their families and go into a *promysel*, a craft. For some the craft was the hunting of fur-bearing animals, and the term *promyshleniki* for fur trappers was born. Interestingly, serfdom and the craft became organically embedded into the concept of hostages, something that entered the Russian psyche, and can be traced through the communist period (when officials were sent abroad, they had to leave their families behind) to current Russian politics, where political loyalty is very frequently based on blackmail, family members included.

The Mongols, sophisticated nomads, created an extremely effective tax system which was proven to be applicable to both nomadic population and settlers. The tax (*yasak*) was payable by the princes of Mongol vassal states, and was a useful model for a system the Russians applied to Siberia when they conquered it. The Mongols accepted *yasak* in grain, military or postal service, horses, saluki dogs and young women, but the preferred payment was fur.

Russia's greatest era of expansion following the overthrow of its Mongol rulers was under Ivan III ('the Great') who managed to unite Russia under Moscow rule and Prince Vasili III, son of Ivan III and father of Ivan IV. The latter, in 1547 as a boy, took the title of Tsar (Tsar Ivan IV), the first Russian Prince to do so. As Ivan IV he was to rule until 1584, a period which saw the consolidation of the Russian state. But Ivan's reign also saw the destruction of his personal guards (the *Streltsy*) and the setting up of the *oprichnina*, a section of the state – about 50 per cent of Moscow Principality, chiefly in the north-east – under Ivan's direct rule. This was controlled by the *oprichniki* (men apart), who instigated a reign of terror, driven on by an increasingly paranoiac Tsar. The *oprichnina* (arguably the first 'official' secret service) was followed by a military campaign against Novgorod (then a centre for foreign trade, and the only democracy in Russian territory). The death toll in the city was probably about 30 per cent of the population, and Ivan was given the name Ivan Grozny. This is usually translated as Ivan the Terrible, which does not seem out of place in view of the purges and their associated cruelties, but which is more correctly translated as 'Dreaded' or 'Fearsome'. During the campaign against Novgorod, a number of Novgordians left the area and settled far away in Siberia, establishing themselves on the Indigirka River, near Arctic Kolyma, near Anadyr (and probably Anadyr itself) and at the Kasilof River in Alaska. The discovery of these settlements in the early twentieth century was sensational, particularly as the people preserved the folklore and customs of Ivan the Terrible's times. However, the Alaskan settlement did not survive, though archaeologists found the remains of twenty-one houses in 1937.

It is likely that early, Slav, migrants into Russia reached the White and Barents sea coasts as early as the eleventh century, but the permanent settlement of Pomoriye, as the coastal region became known (the name derives simply from 'by the sea'), did not occur until the development of the Moscow state. The Pomors would have come into contact with both the Sámi and the Nenets, but this was probably a peaceful coexistence, as it is known that when envoys of Novgorod (which preceded the rise of Moscow) attempted to extract tribute from the Nenets they were killed or forced to retreat several times over a period of some 150 years, implying that the physical ousting of the Nenets from their territory would have been no trivial affair. As noted above, the Pomors may have made early journeys to Svalbard and were responsible for the development of the 'koch', the ship which was to be instrumental in the opening up of Siberia. From the thirteenth century there are tales of northern peoples in Russian writings, which imply trading in furs with peoples near the Ob River, east of the Urals.

Ivan IV's purges of the Russian nobility, his foreign wars which led, eventually, to the weakening of both the *oprichniki* and the state, the devastating raids on outlying Russian regions which the wars generated, together with a prolonged drought which led to famine, brought the country to the verge of bankruptcy. Foreign trade had been weakened by the destruction of Novgorod, but equally catastrophic had been the collapse of the fur trade. At a time when heating was by the burning of wood, homes needed to be of modest size, but the nobility of northern Europe built houses large enough to combine the needs of making an impression with those of a fortress. Impractical to heat effectively, the nobility kept the chills of winter at bay with fur, and the sable of Russia's forests produced the most luxurious of all pelts. Sable were a cornerstone of the Russian economy, but in Russia the animals were now scarce and on their way to local extinction. Desperate for new sources of supply, Ivan allowed the Stroganovs, one of Russia's most powerful mercantile families, to mount an expedition beyond the Urals (though the primary goal of the mission was the stopping of costly raids by the last remaining Mongol Khanate). To lead their invasion the Stroganovs employed a cossack known only by his first and patronymic names, Vasili Timofeyovich, and more commonly called Yermak. Ivan's support for the expedition was ambiguous, and the reprisal raids the incursion into Mongol-controlled lands generated infuriated him. But his anger was quelled by news of the riches in furs that the land beyond the Urals held.

Yermak captured the fortress of Siberia, the capital of the last part of the Mongolian Empire, opening the gates to the vast area of wilderness known nowadays as Siberia. Siberia covers almost 10 per cent of the Earth's land area –

an area into which Europe (excluding European Russia) and the United States, including Alaska, could fit – extending east–west across eight time zones. Polar bears wander its northern boundary, while camels graze on its southern. It has three rivers (Ob, Yenisey and Lena) which each drain basins larger than Western Europe. Lake Baikal is the largest freshwater lake, by volume, on Earth, holding 20 per cent of the world's fresh water: it is also one of the world's most remarkable ecosystems. The list, like the country, goes on.

Siberia's furs attracted fortune seekers who crossed the Urals in numbers. Usually travelling as a group, the *promyshlenniki* – in Siberia the term means a trapper – occasionally protected by cossacks, moved relentlessly east. The mission of the *promyshlenniki* was twofold: to acquire furs and to tax local people or to make them taxable. They operated in much the same way as the privateers of European states, brutal and driven largely by self-interest, and can be viewed in a similar fashion, as either pirates or national heroes.

The tax (*yasak*) collected from the indigenous Siberians was usually paid in furs, mostly sable pelts. In Siberia in the 1630s a sable pelt could be purchased for 3 kopeks. At Moscow's market it could be sold for 20 to 200 roubles, depending on the quality. At the time a Moscow clerk earned 20 roubles annually (a maid to the Tsar would get 50 roubles). So a man bringing twenty sables from Siberia was set up for life: the stakes were high, and, not surprisingly, corruption was rife.

Within twenty years of crossing the Urals the fur trappers had reached the Ob; in twenty more they had reached the Yenisey. By 1630 *promyshlenniki* had reached the Lena. Three years later a group sailed down the Lena to the sea, turned east to reach the Yana and, later, the Indigirka. In 1639 a group of *promyshlenniki* followed the Ulya River to its mouth at the Sea of Okhotsk. It took the cossacks less than sixty years to cross the continent, with no support, military or otherwise, from the government or foreign subsidies. For comparison it took more than 180 years for the Americans to make it overland from the east to the west coast of their land, with governmental support. Conquering Siberia was made without army support (only several small-scale military episodes were known, most of them failures) and mostly by volunteers.

By 1642, cossacks had reached the Kolyma River. Six years later Semen Dezhnev, a former peasant who joined Tsar's service as a cossack in Yakutsk, used the river to begin one of the most astonishing Arctic journeys of all time, one made entirely of his own free will. Together with Fedor Alekseyevich Popov, a manager for the merchant who sponsored the expedition, and ninety men in seven kochs, Dezhnev descended the Kolyma river and turned east. The men were bound for the Anadyr River which, rumour had it, flowed through a country rich in furs to reach a sea where there were walruses in huge numbers, offering the possibility of a fortune both in fur and ivory. Four ships were lost in foul weather early in the journey, a fifth being wrecked on the north Chukotka shore. The ice conditions must have been favourable, however, as Asia's north-eastern point was reached, the remaining two ships turning south with the coast. The two Diomede Islands were sighted, but when the Russians made landfall on the mainland, hostile Chukchis soon made them embark again. One of the remaining ships was lost in a storm, but Dezhnev made it to the Anadyr River. Together with the remnants of his expedition – about twenty-five men in all – he had travelled 3,000km (1,860 miles) in 100 days and had passed through the Bering Strait eighty years before the man after whom it is now known.

Kamchatka was not finally occupied for fifty years after Dezhnev's journey, but by 1700 the Russians had occupied all of Siberia apart from southern Amur, which they lost, by treaty, to the Chinese after initially taking it. Occupied, but not pacified, but as taxation was the aim, occupation was all that was required.

For the settled peoples of the Siberian taiga, the *yasak* system provided the Tsar with a steady income, while the forest provided a steady supply of furs. For the nomadic reindeer people of the tundra, the situation was more problematic: there were far fewer fur-bearing animals, and nomads have no time to check on traps. That there were abuses of the system is hardly surprising; the *promyshlenniki* were tough men accustomed to tolerating the rigours of the Siberian winter, and occasionally their treatment of the native peoples was bad. One Nenet wrote to Tsar Alexei Michailovich (who reigned from 1645–76) describing the horrors inflicted on his people by Russian incomers:

> For five years...they have been robbing and abusing us in every way, taking from us with violence and all manner of threats our sable and beaver furs, our deerskin bedding and clothing, our ropes and all kinds of footwear, our geese, ducks and reindeer flesh...We are left naked and barefoot and shall perish completely.

The phrase 'naked and barefoot' must not be taken literally – it is poetic (thyming in Russian) and cites the Russian Gospels – but is an indication of the harsh treatment the Nenets received. However, what is equally important about the letter is that the Nenet was literate, and that he felt able to write directly to the Tsar. In British-controlled Canada two centuries later it is unlikely that a native chief could have written such a letter or that it would have been allowed to reach the king.

In an attempt to improve the position, *voyevodas* (governors) were appointed to collect *yasak* rather than leaving it to the *promyshlenniki.* But the appointment of suitable people was difficult. The *voyevodas* could be (and probably needed to be) as ruthless as the trappers they were overseeing and the condition of the native peoples often did not improve. Indeed, it sometimes became worse. The governors were installed in forts (*ostrogs*), usually on rivers as these were the most efficient way to travel (and offered the possibility of catching fish which was handy if supplies became depleted). These forts often became the target for native resentment, and many were attacked over the years, their 'garrisons' being killed. Such attacks brought retribution, often terrible in its cruelty and extent. Ultimately the native peoples of those areas of western and central Siberia that were rich in fur-bearing animals were pacified, though occasional flashes of resistance persisted for several decades.

That the native peoples were not always friendly innocents driven to violence by the treatment meted out to them is true. The most aggressive nation in the North were the Chukchi. They were able to coordinate substantial military attacks and had kept their neighbours at bay for years before the arrival of the *promyshlenniki.* The first time the Russians encountered Chukchi aggression was in 1641, when a group of *promyshlenniki* led by Ivan Stadukhin was robbed on the Kolyma River. Dezhnev's expedition had also fought a brief battle with Chukchis before the sixth ship of the fleet, that captained by Popov, was driven onto the eastern Chukotka shore by a storm. Discovered by Koryaks, many of the Russians were killed (though some escaped), one survivor being Popov's Yakut mistress, probably allowed to live because of her sex and held captive as a trophy of war. Ironically, it was Dezhnev's chance discovery of the woman when he later encountered the Koryak group – with more men, all of them better able to look after themselves than a shipwrecked crew – that allowed him to discover Popov's fate. Years later, ruins found on Kamchatka and the story of Koryaks there revealed a final twist: that the surviving crew members had escaped from Chukotka and made their way to Kamchatka where their dramatic arrival had caused them to be revered as gods until their behaviour proved far less god-like and they had been killed. Dezhnev managed to establish tolerant, even friendly, relations with the Koryak he met, but these were disrupted by the arrival of another group of *promyshlenniki,* who treated the native population appallingly. Eventually the Koryaks revolted, killing many of the new group, something which forced Dezhnev's group to retaliate.

The history of Russia's interaction with, and subjugation of, the native peoples is essentially the history of the fur trade. The subjugation is not portrayed as a form of ethnic cleansing, but in reality there was nothing to compare with the massive population decreases seen among the native peoples of America. Indeed, many populations had increased by the nineteenth century. However, it is, of course, true that the introduction of strange diseases took a toll. The Russians also took advantage of the enmity of many of the native groups with their neighbours. As with the native peoples of the Nearctic, where the Inuit fought with the Indian tribes to the south whenever they met, and the Indian tribes fought each other, Eurasia's Arctic dwellers also routinely fought with each other. The occasionally expressed view that the native peoples of the Arctic lived in harmony both with each other and with nature is at odds with reality. They exhibited the same sins as the 'civilized' people who first met them and would undoubtedly have ravaged the landscape in much the same way if only they had had the technology and resources to do so. Russian exploitation of the natural enmity of the northern tribes often resulted in an increase in inter-tribal conflict, something which is absolutely contrary to the often-seen suggestion that the Russians brought stability to the region. Group migrations, caused by a desire to escape the taxation and occasional brutality of Tsar rule, also added to inter-tribal tensions and undoubtedly created more problems. Ironically, even when faced with a common enemy, the native groups mostly did not unite. The fiercest resistance the Russians encountered was against the Chukchi and Koryak in Chukotka. Although there were occasional periods of unity of purpose between the two groups, in general, as soon as the Russians had departed to gather breath, the Chukchi and Koryak would resume fighting, a policy that weakened them in readiness for the Russians' return. By the time the Koryaks had finally been subdued their population had been reduced to no more than 40 per cent of the total at first contact (as far as tax-paying males are concerned).

For many of the native groups of southern Siberia, the *yasak* demands of the Russians amounted to nothing more than a change of ruler, as the groups had been paying tributes to the Mongols for many years. To the north, those groups who had escaped such tributes were less willing to accept Russian demands and needed harsher tactics to bring them into line. When the Russians first moved south from the Anadyr River area to explore Kamchatka they were attacked by both Koryaks and Itelmen. The *promyshlenniki* suffered losses as the native groups fought a guerilla-type campaign, using skis and dog sleds to aid speedy attacks and rapid retreats. Ultimately the essentially 'stone-age' weaponry of the natives was no match for the Russian weapons; when the Russians found native villages they exacted a terrible revenge, burning tents and possessions and slaughtering everyone they caught. Further south in Kamchatka the Russians again used their divisive tactics, siding with one group of Itelmen against a rival group, even mounting a raid of their own to prove friendship with the first group. The tactics worked, and the grateful Itelmen willingly offered a tribute in furs. At the southern tip of Kamchatka the Russians learned of the existence of the Kuril Islands, and heard that beyond them there was a land where people lived in stone buildings and had possessions that the Itelmen could only dream of owning. The Russians were puzzled: it would be some time before they realized that what was being described was Japan.

Russian control of Kamchatka was tenuous at first, the harsh treatment of the locals causing several uprisings in which cossacks were killed, followed by savage reprisals that often involved the massacre of many and the forced migration of villages to sites where the people were easier to control. The *promyshlenniki* also robbed the Itelmen, taking not only furs, but boys and girls who were used as, or sold as, slaves. In many ways the position of the Itelmen was worse than that of other native peoples. It seemed that the further away the *promyshlenniki* were from Moscow, the more they behaved as despots, with the native peoples the butt of their depraved advances. Kamchatka was supplied across, and furs were carried away across, the Sea of Okhotsk: it was, to all intents and purposes, an island. This resulted in a sense of isolation that allowed the growth of tyrannical governors and inhuman treatment. On the Aleutians, which literally were islands, the treatment of the Aleuts was even worse. The population of Itelmen, reduced by disease, maltreatment, death in rebellion, suicide and despair, probably fell to below 2,000, a reduction of perhaps 90 per cent. At the same time, security fears over the fur trade in the Bering Sea led to an influx of military personnel to Kamchatka. Together with the fur trappers, the fur traders and the trade officials, the Russian population grew rapidly so that émigrés eventually outnumbered the locals.

In northern Kamchatka and southern Chukotka, the Koryaks were even more of a handful than the Itelmen, constantly attacking Russian groups. Eventually the Russians were forced to build forts as brutal retaliatory strikes failed to achieve submission. The Koryaks responded by attacking the forts, burning several to the ground. Currently they are the dominant indigenous people in Kamchatka and the Magadan region, totalling about 9,000 people.

North again, the Chukchi proved even fiercer opponents, utterly fearless in the face of death. As there were no fur-bearing animals in Chukchi lands, the Russians left them alone for several decades, invading again only when the decision was made to bring all the land of north-east Siberia under Russian control. Subjugation came at the expense of considerable bloodshed on both sides. In one battle in 1747 a force of about 600 Chukchi surrounded a regiment of the

Russian army numbering about 140, led by Major Pavlutskiy, a Byelorussian, and wiped them out. Pavlutskiy had actually been sent to protect the Koryak tribes and their reindeer, and the slaughter, close to the Anadyr fort, which was robbed, was a scar on the army's dignity. The Anadyr fort was closed in 1771, an indication of Russian defeat. As a result the Koryaks had to move south to Gizhiga and Kamchatka, while the Yukagirs were forced to move to the Kolyma, vacating huge areas. Compelled to take real action, Moscow dispatched an army column which inflicted significant damage on the Chukchi, but failed to achieve total submission. Finally, in 1776, Catherine the Great used the diplomatic talents of Chukchi-born Nikolay Daurking and Timofey Shmalev, *voyevoda* of the Gizhiga fortress, to bring the Chukchi to the negotiating table. The Chukchi were forced to accept subjugation, but were tax exempt for ten years. Legislation of 1822 gave the Chukchi the right to fix their own tax rate, while in 1885 a governmental inspector was reporting that 'the Chukchi Peninsula is governed by itself and does not recognize any other power'.

In 1676 Tsar Alexei died. He was succeeded by his eldest son Fyodor, but Fyodor died childless in 1682 and Peter, Alexei's son from a second marriage, was proclaimed Tsar in preference to his older half-brother Ivan (Fyodor's younger brother), as the latter was deemed mentally incapable. Peter was only ten years old, and for many years his supporters had to contend with the machinations of his much older half-sister Sophia, who favoured Ivan as a way of maintaining her own position. Only in 1694 was Peter able to take full control of the reins of power. He proved an able ruler, justifying the addition of 'the Great' to his name. He built a new capital, St Petersburg, and developed Russia along European lines. He also fought a long war against the Swedes. Such things cost money and Peter was soon looking for ways to increase his exchequer. Inevitably his eyes turned eastwards. Siberia had provided untold riches in furs, but now it had to provide more. For it to do so, Peter needed to improve communications with the far east of his realm. More importantly he needed to expand his sphere of influence to include, if possible, new lands which might provide yet more furs and, perhaps, other resources. In Chukotka the Russians had heard the locals tell of lands to the east. These lands had not been visited often, but the Chukchi who had reached them had returned with furs of an animal that had a curious black and red striped tail. The Russians reasoned that the lands were either part of America or islands between Asia and America.

Tsar Peter therefore decided to map Siberia's coastline, the better to understand its geography, and to send ships east from Kamchatka. The mission of the latter was to discover whether Asia and America were actually joined – the memory of Semen Dezhnev's voyage, which had already answered that question, having been forgotten with the passage of time. The mapping of Siberia became known as the Great Northern Expedition and was remarkable for the persistence of those who undertook it (though the penalty for failure almost certainly contributed to this: leaders who did not maintain proscribed schedules were often court-martialled and demoted, sometimes to the lowest service rank, even if their excuse was that ice had frozen-in their ships, or cold and rugged terrain slowed their progress). In a little over ten years the whole northern coast of Russia, together with the eastern coasts of Chukotka and Kamchatka, had been surveyed. It was an astonishing achievement.

To establish the second part of Peter the Great's mission, the Danish-born Vitus Bering sailed north from Avacha Bay in southern Kamchatka, discovering St Lawrence Island and passing through the strait that now bears his name (Bering's name was given to the feature – and to the sea south of the strait – by James Cook, who was unaware of Dezhnev's voyage. The name has stuck, Dezhnev's name only having been attached to the cape which is the easternmost point of Eurasia.) On this first voyage Bering did not turn east at the strait and so missed discovering Alaska. He also failed to turn west and so did not definitely prove that Asia and America were not joined. It was almost 100 years before it was proved beyond doubt that there was no connection between Alaska and the land the Chukchis knew existed off Chukotka's northern coast; the land the Chukchis occasionally glimpsed was Wrangel Island.

Despite Bering's failure to answer the question of a land connection between Asia and America he was given command of a second expedition charged with finding out exactly what land lay across the sea east of Kamchatka. It was on this second expedition that Alaska was discovered. The expedition ended in tragedy, however, as Bering and many of his crew died of scurvy on what is now named Bering Island, one the Commander Islands (which were also named after Bering). The remnants of Bering's expedition carried the news of Alaska's discovery back to Kamchatka. They also brought the pelts of sea otters and started a second rush for furs that would only end when the sea otter and the northern fur seal had been brought to the brink of extinction. The sea otters were found on the Aleutian Islands, which became the target of the *promyshlenniki.* Russian operations in the Aleutian Islands had two stages: the first involved visits by *promyshlenniki,* a chaotic period based on a first-come-first-serve basis; the second wave came under the management of the Russian-American Company (ROC). The first wave lasted from the arrival of the remnant of Bering's second expedition in Petropavlosk to 1799, the year of the establishment of the ROC. It was a time of general lawlessness and undoubtedly there were atrocities against the Aleut population of the islands, but these have to be seen in the context of the times. The Russians did not displace the Aleuts, and after the arrival of Aleksander Baranov things improved greatly; schools were built for them and some Aleuts were sent to schools in Japan to learn navigation and other skills. (The first school was opened in 1794, for both Aleuts and Russians, by Ioann Popov Veniaminov (1797–1879), who also created an alphabet for the Aleuts and translated the Bible into Aleutian.) One of the reasons so much is known of the atrocities carried out by Russians against the Aleuts is from letters of protest sent to the Tsar, indicating literacy. In general, the treatment of individuals was probably no worse than that of Hudson's Bay Company men against native Canadians. The history of British and American treatment of non-Arctic native peoples also makes Russian treatment of the Aleuts look somewhat less horrific. The Americans' treatment of the First Nations people of the Lower 48 States amounted to an (almost successful) ethnic cleansing; at the same time the Americans almost wiped out the population of North American bison. Neither is the British and American slave trade an example of the enlightened treatment of a native population.

On the islands the wildlife also suffered. When they had reduced the population of sea otters to the extent that hunting them was uneconomic, the Russians imported Arctic foxes, setting them free on uninhabited islands. Having evolved in the absence of terrestrial predators, ground-nesting birds were easy targets for the foxes: only three islands, their cliffs too rugged for easy landing, escaped the introductions. Endemic sub-species clung to life on those, though fox eradication programmes have now ensured the survival of several critically endangered species. For one species there was to be no happy ending. Steller's sea cow – named after Georg W. Steller, the naturalist who accompanied Bering's second expedition – was hunted to extinction within twenty-seven years of Steller first describing it, its fate sealed by a lethal combination of docility, slowness and being good eating, which made it the immediate target of the fur trappers.

Early in the conquest of Siberia, Christian missionaries had moved in to the more accessible, western, section. Their attitude to the native Siberians often differed little from that of the trappers, with conversions being beaten out of those who did not choose voluntary baptism. There are tales of forcible baptism, even of people being tied to poles which were plunged into streams and lakes (sometimes through holes cut in the ice), a procedure which, not surprisingly, did not fill the sufferers with great enthusiasm for the new religion. But these early attempts at evangelizing the native groups created conflict between the trappers and the missionaries, and between the State and the missionaries. On the bright side the missionaries were recording local languages, making dictionaries and setting up schools.

By the nineteenth century the situation had changed. There had been an influx of peasant farmers to the more agriculturally suitable areas of Siberia and the largely nomadic native groups of the north had been decreed to be settled, a decree that allowed them to be taxed in the same way as the truly settled peasantry to the south, and by royal decree *yasak* was abolished in 1822. As the level of taxes were, in general, set substantially higher than the *yasak* had been, loss of *yasak* income was now less of a problem. Now not only could the church continue with its missionary work, but the ban on forced conversion was lifted (or, at least, ignored). The first missionaries were enthusiastic baptizers and keen to destroy any semblance of the old religion of the native groups. All the northern dwellers had belief systems which can be gathered under the heading 'shamanistic', the belief in a spirit world to which the group shaman could travel to intercede on behalf of them. The shaman usually 'travelled' by entering a trance-like state induced by the rhythmic beating of his drum. Although ritual objects were few in number, and ritual sites fewer still, the Christian missionaries sought them out and destroyed them. Those that accepted baptism invariably had little understanding of Christianity and forced conversion did not disrupt native group lifestyles, as it was customary to add the Christian God into the pantheon of the shamanistic gods. Often, the shamanistic rituals in the yurts of Chukchi or Nenets were performed before an Orthodox icon.

Following the Revolution the communists showed no more enthusiasm than the Tsarist regime for allowing the northern peoples any true autonomy or complete exemption from the privilege of being ruled by Moscow. While the Committee for the Assistance to the Peoples of the Northern Borderlands (the lengthy title usually shortened to Committee of the North) set up in 1924 (the year of Lenin's death, and Stalin's promotion) had, as its primary aim, the integration of the northern peoples to the socialist state, it did seek to regulate trade so that the worst excesses of the past were eliminated. The Committee also had several enlightened ideas, such as the production of primary school textbooks in native languages, and the introduction of elements of self-governing. In 1930 the Chukchi, Koryak and Evensk autonomous districts were formed, thus making the Committee obsolete, and it quietly died in 1935. In 1930 in Leningrad a new Pedagogical University was opened in order to prepare teachers for schools in the Arctic. It is still active, now as a branch of the Gertzen Pedagogical University in St Petersburg.

However, Stalinist Russia had no time for those wanting to live a life outside its strictures. The reindeer herds of the far northern dwellers were subject to the same collectivization as all agricultural holdings, and with bulk sales of meat only allowed through the *kokhoz*. Local shops were also under state control, and in the Arctic were largely dependant on supply by sea along the North-Eastern Passage, which from 1930 was governed by the Sevmorput, a Committee within the framework of the Council of Ministers.

The Koryak resisted collectivization, killing their animals rather than allowing them to become State property, although this gesture did not halt the imposition of the collective. Some Nenets also avoided collectivization, formed an organization called 'Mandalada' and revolted in 1932 under the leadership of a shaman called Barkhatov. In one episode the Nenets took party officials hostage and attacked a Red Army regiment. Eventually they killed the hostages. In the same year the town of Khatanga was captured by Nenets and Dolgans, with all 100 Russian residents being taken hostage. In the autumn the two peoples sent a letter to Moscow asking for the unlawful taxation and despotism of local authorities to be ended. They also sent a telegram to the 'Peoples of the World'. In 1934 there was a further revolt. These rebellions were not against collectivization as such, but against the monetary tax imposed on the native populations. The outcry reached such proportions that the authorities reviewed the question of taxation and agreed to pay the money back; this was actually paid in 1944, despite the Soviet Union being involved in a bitter and costly war. However, some of the Nenet rebels were sent to the camps (although they were hardly alone in this fate during the period of Stalin's control).

The position of the Koryaks and Chukchi differed from that of the native groups to the west because Russian rule in Siberia's far north-east had always been tenuous. During the late nineteenth century American whalers chasing bowheads in the Bering and Chukchi seas frequently visited Chukchi villages where they bartered American-made goods and alcohol for furs and ivory. The trade was so prevalent that it is said that in many villages all manufactured goods were American as they were better quality and cheaper than Russian equivalents, and included items the Russians did not make. The Japanese and British also traded with the Siberian far north-east. Although Russia resented the trade it is likely that without it the situation for the peoples of the far north-east would have been dire (in part, it has to be said, because of the ruthless exploitation of the bowhead and walrus population of the northern Bering Sea by the very nations who were trading with the Chukchi).

When Russia finally got a grip on the Siberian north-east they took over the sea mammal hunts and continued the over-exploitation. They forcibly removed the Yuppiat from their coastal villages (the fear being not so much that they would reduce the available catch as that they might otherwise escape to Alaska), relocating them inland where their skills were useless and their presence caused friction with the Chukchi. Forced to take on the jobs which no one else wanted at very low wages many Yuppiat lived in abject poverty.

Three other issues added to the overall misery of the northern peoples. The first was the wholesale exile of so-called dissidents and criminals to Siberia during Stalin's purges. Siberia had been a sink for enemies of the State and criminals since Tsarist times, but the rate of exile increased hugely. Some ended up in corrective labour camps (the infamous gulag). The second was the removal of native children from their parents for education in places remote from them. Here, the children were educated in the ideology of the State in a bid to integrate them into mainstream communist Soviet society. As with the ill-fated Indian boarding schools in the United States, and the Canadian residential school system, the idea was the focus of international criticism. The practice was suspended in 1990, but its absence has lead to an illiteracy problem across Chukotka. Ironically, many Chukchi see the Stalin period as a golden age, better than the present time. Thirdly, when, at the end of the 1950s, the era of the gulags was finally over, an awareness of the potential mineral wealth of Siberia led to a further influx of Russians into native lands, while the Cold War resulted in the military occupation of many northern areas.

After his death, in 1953, the rigid command society of Stalin's period was replaced by what is perhaps best described as incompetent paternalism; the continued industrial development of the Arctic was tempered by a number of inefficient and often useless aid programmes. In another irony, during the *perestroika* period the central authorities considered these the times when the northern native peoples were 'closest to heaven', an idea utterly at odds with the view of the peoples themselves.

As communism failed, food shortages and the inability to pay workers' wages led to hardships across Russia, though in the north these were exacerbated by the loss of professionals who had 'emigrated' to Siberia, but now chose or were forced to return to European Russia in the hope of a better life. Some also left because their native lands had become separate countries. So many years after the effective collapse of their own societies, without the safety net that the Soviet system had offered before Russia's economic decline, the native populations suffered badly.

In 1990, just a few months before the break-up of the USSR, the Russian Association of Minority Peoples of the North (RAIPON) was founded with the aim of addressing these problems. Today the organization has expanded to cover forty-one indigenous groups, a total of about 250,000 people occupying some 65 per cent of Russia's land area. However, RAIPON has not been entirely successful in improving the lot of Arctic dwellers (or those of other areas of Russia) as a succession of laws aimed at giving enhanced legal status to the people and their land claims in the early 1990s was vetoed by Boris Yeltsin, after which the enthusiasm of central government for the project waned. In Russia, as elsewhere, those who have a vested interest in ensuring that the rights of native people remain minimal argue that offering anything that smacks of enhanced rights to native groups is counter to the idea of equality for all citizens. While such arguments are repeated and discussed *ad nauseum,* rampant capitalism by financial oligarchs continues apace.

Apart from the southern island of Novaya Zemlya, Russia's Arctic islands were not inhabited in modern times (though Wrangel and Zhokhova islands were inhabited by Paleolithic hunters) and were discovered only after the mapping of the mainland and the arrogation of the native peoples of Siberia's north had been completed. After Barents wintered on Novaya Zemlya's northern island in 1596–7 that island group became reasonably well known. Novosibirskiye Ostrova (the New Siberian Islands) were first recorded in 1712 when the fur trappers Permyakov and Vagin on the mainland noticed a herd of reindeer heading south towards them across the sea ice. They reasoned that the herd must be coming from land and followed their tracks to two islands, from which further tracks headed further north. In 1770 the Russian *promyshlennik* Lyakhov made the first description of the island now named after him. In 1773 the same trapper discovered a larger island to the north. On this he found a copper kettle, indicating that more discreet trappers or hunters had come this way before; the island is still called Kettle Island.

The Chukchi had told early Russian travellers about the land they occasionally saw off Chukotka's northern coast, but it was not until American whalers began to operate in the Chukchi Sea in the mid-nineteenth century that it was formally discovered. In 1848 Henry Kellet, captain of the *Herald,* discovered Herald Island, naming it after his ship. From the island he saw another to the west, naming it Kellett's Land, though the name was changed to Wrangel in 1867 by Thomas Long, another American whaler, either unaware of Kellet's claim or preferring a more historical name: Baron von Wrangell (by convention, following Long's mistake, the island is spelled with a single 'l'). Wrangell's 1820–3 expedition had explored the north-east Siberian mainland, but had failed to spot the island despite also being told that it could be seen on clear days. In 1921 Vilhjalmur Stefansson (Canadian-born of Icelandic parents) made an ill-fated attempt to settle the island. Stefansson had already led an expedition in 1913–14 that had ended disastrously after the *Karluk* became entombed in ice, forcing the crew who had remained on board to abandon her in favour of crossing the ice to Wrangel and Herald islands. Stefansson's attempt to settle Wrangel led the Soviet Union to mount an expedition in 1924 that formally claimed sovereignty of Wrangel. In the same year an American expedition landed on Herald Island to claim it as American territory. They found the remains of the four men from the *Karluk* who had made landfall, but died, probably of exposure, on one of the Arctic's most inhospitable islands. The 1924 American claim to sovereignty was preceded by a Russian claim in 1916 (although there had been no Russian landing at that time). The United States has never formally claimed the island and supports the Russian claim.

Franz Josef Land was formally discovered by an Austrian expedition in 1872, though as with Wrangel the archipelago had almost certainly been seen before. Following its discovery the archipelago was used for several expeditions seeking to reach the North Pole, and in 1901 was surveyed for the first time by a Russian expedition in the ice-breaker *Yermak*. Sovereignty remained unclear until 1926 when the Soviet Union made formal claim, prompted by a half-hearted British attempt on the Pole from Franz Josef in 1925. The 1926 Soviet Union claim may also have been prompted by an earlier Canadian claim to all the islands within its sector of the Arctic, as the Soviets used the same terminology (thus claiming anything so far undiscovered). Yet despite the claim Norwegian hunters continued to operate in the archipelago: only in 1930 did the Soviets turn Norwegian ships away and, after establishing a weather station, formally raise the flag of the USSR above Hooker Island. The Norwegians, nervous over their own claim to Svalbard, expressed annoyance over the claim, but as no other government took any notice, Franz Josef Land became part of the Soviet Union.

The final Russian Arctic archipelago, Severnaya Zemlya, was not discovered until the Russian Arctic Ocean Hydrographic Expedition of 1910–15, when two ice-breakers, *Taymir* and *Vaygach*, exploring and more thoroughly mapping the northern mainland under the command of Admiral Kolchak, made the discovery. The ice-breakers were also making the first east–west transit of the North-East Passage. The first transit, from west to east, had been made in the ship *Vega* commanded by Adolf Erik Nordenskiöld in 1878–9.

Although, as noted earlier, Russia's Arctic islands were largely uninhabited, the USSR did successfully establish scientific/exploration camps which have been more or less maintained, thus establishing ownership in international law. In 1926 the Presidium of the Supreme Soviet decreed that all lands between the Russian northern coast and the North Pole not already 'the territory of any foreign state are declared territory of the [USSR].' In passing this decree the Soviet Union was, though not as precisely, invoking the 'sector' approach to Arctic ownership which had already been set down by the Canadians (see page 87).

Alaska

The discovery of Alaska by Bering's second expedition, and the subsequent subjugation of the Aleuts, have already been considered above. But Russian exploration of Alaska did not stop at the Aleutian Islands. Within forty years of the first landing, Russian fur trappers were hunting sea otters in the waters of Cook Inlet.

At first ships arrived for the hunting season, but then Gavriil Pribilof discovered the islands which now bear his name. On them he found millions of northern fur seals, and the Russians transported Aleuts to the two islands of St Paul and St George in order to hunt the animals. With the discovery of the islands, it became clear the idea of sending ships for the hunting season alone would no longer work – a permanent base was required. In 1799 the era of chaotic, unorganized expeditions by private ships ended and Russian expansion into Alaska began a new phase. The Russian-American Company (ROC), privately owned by Irkutsk merchants, was given a hunting monopoly in the new territory by a decree of the Tsar Pavel I. The company was modelled on the British East India Company and operated the sea ports of Okhtotsk, Petropavlovsk-Kamchatskiy and Ayan. By 1800 the company was so important to the Russian economy, selling furs principally to China, that the headquarters were moved to St Petersburg (then Russia's capital), despite Irkutsk being closer to both the furs and China.

In Alaska, Aleksander Baranov was in charge, having been appointed in 1794, and the new territory had a Russian population of perhaps 700, many at the capital of Novo-Arkhangelsk (now Sitka), which was established in 1799. Under Baranov's astute leadership Russia expanded its holding in North America. The expansion was largely peaceful, though there were conflicts between the Russians and Athabaskan Indians, particularly the Tlingit. One battle, in 1804, close to Sitka, involved men from a single Russian naval vessel, the only time Russian military personnel were ever to set foot in Alaska. As well as exploring and settling south-eastern Alaska, the Russians also went north, Otto von Kotzebue reaching the north-western Yuppiat village which now bears his name. Russians also set up trading posts in Bodega Bay, California, not far from San Francisco. Baranov, having spent twenty-seven years with the ROC, and having waited years for a replacement, finally retired in 1819: he died on his way home to Russia just a few weeks short of his seventy-second birthday. His body was tossed into the sea near Java. He has no grave, or tombstone and left behind one suitcase, a couple of souvenirs and two shares in the Russian American Company. When in 1820, the ROC monopoly was up for review, the financial check of the company records revealed profits almost double those anticipated: Baranov had never attempted to extract anything from the company that was not due to him. Ironically, after the review, with new governing rules in place (and Baranov gone of course) the ROC went into decline. Interestingly, as part of the new rules, all native peoples in Alaska were to be treated as subjects of Russia, which meant that the collection of *yasak* was banned.

By 1820 exploration had brought Russians into contact with the Americans of the United States and the British in Canada and it was clear that Alaska's borders needed to be defined to avoid conflict with each. In 1821 Tsar Alexander I banned foreign ships from Pacific waters north of 51°N, with a threat of confiscation for any that intruded. As both Britain and the United States laid claim to land north of that line the ban was vigorously opposed, not least because each claimed that Russia's position in Alaska was that of a trading company with outposts rather than a true settlement. It was a fair point: at the time the total number of ethnic Russians in North America was probably less than 1,000 and the documents delineating Alaska's border were actually the first statement of any land claim. The ROC had been forbidden to set up inland trading posts, all Russian activity being along the coast. Moreover, the long list of titles of the Tsars (or Empresses) of Russia never included anything such as 'Prince of Alaska'. There were no Russian naval bases, soldiers or border guards, and no surveyors had ever been employed; Russian occupation of Alaska was essentially a private enterprise of Irkutsk merchants, occasionally with official support, and that offered only with political strings attached.

Both Britain and the United States called for negotiations to define the borders of Russian America. In 1824 the US agreed that Russian holdings in Alaska would extend only as far south as 54°40′N and the Russian Californian bases would be withdrawn. The curious latitude was chosen because of the discovery, in 1774, of relics which are credibly assumed to be from Aleksei Chirikov's men. Chirikov commanded the second ship of Bering's 1741 expedition and is known to have spotted the Alexander Archipelago. The relics were discovered the southern end of Prince of Wales Island, at approximately 54°40′N.

Negotiations between the Russians and the British were less straightforward, particularly as British searches for the North-West Passage had already brought them as far west as Prudhoe Bay (and as far east as Point Barrow from the Pacific Ocean). Negotiations started with the Russians claiming all land to 139°W, while the British claimed all of south-eastern Alaska as far as 59°N, basing their claim on the explorations of James Cook and George Vancouver. A compromise was reached, with the British agreeing that the 54°40' latitude represented the southern tip of Alaska in exchange for the Russians agreeing to move the western border to 141°W. All that remained was to agree the demarcation where the 141°W line of longitude reached the southern coast. The Russians had become increasingly concerned that the Hudson's Bay Company was encroaching on its fur trade with the North American interior and felt that they needed a section of mainland (the *lisière* as it was called) close to the Alexander Archipelago to maintain some foothold in the trade. Without it, they feared, trade on the archipelago would wither away and the advantages of holding it would die with it: the British would then gain control of south-east Alaska. Again a compromise was reached: where the 141°W longitude reached the mountain range which was believed to run continuously along the coast (the range is still called the Coast Mountains) the border between Alaska and Canada would follow the mountain crest, with the proviso that at no point would the border lie more than 10 marine leagues from the sea (1 marine league = 1/20 latitude degree = 3 nautical miles = 5.6km). The agreement, finalized in 1825, also included a provision (Article VI) which granted British subjects the right of transit from Canada to the Pacific Ocean 'in perpetuity'. This right was to cause problems when Alaska became American.

But Alaska was to remain Russian for a further forty years. This period saw the beginning of the settlements of Ninilchik and Seldovia on the Kenai Peninsula, as Russians employed by the Russian American Company chose to remain in Alaska rather than return home, and an increasing influence of the Russian Orthodox Church in the lives of native Alaskans, as missionaries spread out from Sitka and other church strongholds. Today, Russian churches are a familiar sight to travellers in southern Alaska. But as well as seeing an increase in the Russian development of Alaska, the period also saw an increase in conflict with both the native people, the British and the Americans. Native groups attacked several settlements, killing Russian and Aleut inhabitants, and an entire exploratory group in north-west Alaska was killed by outraged natives. Hudson's Bay Company ships attempted to use south-eastern Alaska waters to set up a trading post (on British territory), and several were confronted and turned around. American traders had been a problem for many years (even during Baranov's time), not only illicitly trading furs, but also exploiting Indian inter-tribe rivalries by trading guns for furs, then buying the captives the guns helped procure as slaves. To add to this problem American (and French) whalers began to hunt right whales in the Gulf of Alaska, causing an angry exchange of correspondence between the respective governments, particularly as the American whalers often behaved appallingly, landing at Russian outposts where Aleuts had dried fish and collected wood for winter, and stealing these supplies. Within ten years the Americans were also whaling in the Beaufort and Chukchi seas, though at least on Alaska's northern shores there were no Russian outposts to plunder. In south-east Alaska the Russians were forced to allow the Hudson's Bay Company to set up bases on their territory in exchange for supplies as the distance from mainland Russia and the dangerous Bering Sea – about half of all the ships taking supplies to Russian America floundered along the way – made deliveries erratic.

In 1825 Tsar Alexander I died, and Nikolas I became Tsar, although his ascent to the throne was overshadowed by the Decembrists revolt aimed at limiting the rights of the monarchy. The revolt was defeated. One of the five ringleaders hanged was K. Ryleev, the chief of the ROC office in St Petersburg, something which reduced the company's standing with officialdom.

Then in 1854 the Crimean War broke out. The Russians of south-east Alaska and Hudson's Bay Company personnel agreed to remain neutral, but the conflict brought home to the Russian government just how far away Alaska was, and how difficult it would be if they were required to defend it. The British and French fleet attacked sea ports of Ayan in the sea of Okhotsk and Petropavlovsk-Kamchatskiy. Ayan was set on fire, newly built ships were destroyed or captured, and the townspeople fled to the forest. General Zavoyko, Commander of Petropavlovsk, alerted by the Hawaiian King about the imminent attack, put up stout resistance, aided by local people – Russians, Aleuts and Kamchadals. In the battle, which took place both on land and sea, the Russians beat off the attack and the British-French navy retreated with two captured ships.

The Crimean War ended in defeat for Russia. The war had been expensive, and the victorious British were at the border of Alaska. The Russians had built forts which usually provided adequate protection against the local native groups, but they were usually close to the sea and the defenders knew that a cannonade from a British naval vessel would rapidly reduce every fort to rubble. If the British chose to take Alaska there was little Russia could do to stop them; then only the Bering Sea would stand between them and Siberia. Russia would have its most feared rival to both east and west. To make matters worse, the American cod fleet, sailing out of San Francisco, was fishing the waters of south-east Alaska. It would be much better to sell Alaska to the Americans than to fight them over fish and, at the same time, to acquire a buffer between Siberia and British Canada.

The sheer size of Siberia has already been alluded to, but in an era when transport was both limited and slow, the physical size of the region translated into the time it took to cross. In the late eighteenth century the story (claimed to be true – and perhaps it was) was told of a group of six young Itelmen women who were to be presented at court in St Petersburg. Each was provided with a chaperon, a military officer of impeccable character. The six set off from Kamchatka; by the time they had reached St Petersburg each had given birth to not one but two children fathered by their chaperons. By the mid-eighteenth century, the fastest way to reach St Petersburg from Kamchatka was to sail south-east to California, cross the United States by rail, then the Atlantic by ship. Once Europe was reached the railways could again be used, though it was said that even if the traveller used a horse from his French port of disembarkation he would still reach St Petersburg many weeks ahead of someone travelling west across Siberia.

Negotiations therefore began with a view to selling Alaska to the United States, but they were soon stalled by the outbreak of the American Civil War. After an initial false start they began again with the cessation of hostilities, the negotiators being the Russian Privy Councillor, Envoy Extraordinaire to the United States, who preferred the Western rendering of his name – Edouard de Stoeckl – and the US Secretary of State, William Seward. Seward had been a champion of the purchase for some time. He was also in favour of the US annexing the Hawaiian islands and buying Cuba (though he was a prominent critic of slavery), Greenland and Iceland so as to ensure a sufficient buffer zone to any potentially hostile country. However, what every US citizen and politician would now unquestionably see as one of the great deals of all time was not seen as such at the time. Seward's purchase was ridiculed as Seward's Folly, Seward's Icebox or Walrussia. On 18 March 1867 Russia's land in North America was sold for $7,200,000, a price which worked out at 2 cents per acre (and was therefore a better deal than the purchase of Louisiana from the French in 1803 which had cost 3 cents per acre). The Russian-American Company was sold for an unknown price to a company in San Francisco, which continued trading under the name of the Alaska Commercial Company (or Northern Commercial Company). The company rented the Commander Islands and the Pribilof Islands and continued its fur operations much as it had done before, with the same personnel. It is believed that this company made huge profits during the Alaska Gold Rush.

The Alaska Purchase Agreement downgraded the native population from subjects of the Russian crown to 'uncivilized native tribes', subject to the US regulations regarding aboriginal tribes. Russians were given the choice of either returning to Russia within three years, or becoming citizens of the United States.

Despite the Agreement, it was to be seven months before Russia formally handed over control at a ceremony at Sitka. On 18 October 1867, 100 Russian and 250 American soldiers lined up outside the governor's house of the Russian-American Company and the Russian flag was hauled down for the last time. Interestingly, the ROC had never been allowed to sail under the Russian flag, and had a flag of its own, but for the ceremony Americans demanded the Russian flag be used. A freezing rain was falling and the flag stuck to the pole, almost as a protest against the occasion. After much tugging a Russian was ordered to climb the pole and release the flag. The two, man and flag, slid down, the Stars and Stripes were raised and Russian America had become part of the United States. If the delay between signature and flag lowering seems long, it was actually short if measured against the time it took for the Russians to receive the money they were owed, the US politicians wrangling for many months and payment only being made amid accusations of bribery and embezzlement. Edouard de Stoeckl claimed a fee from the US Congress. He never went back to Russia, and died in France.

One thing remained to be done. The new colony required a name. Although in the brief history above Alaska has been given the name that is now so familiar, the Russians had not used that name before 1853, and the agreed sale had been of the Russian holdings in America. The name chosen, Alaska, is said to derive from the Athapaskan *Alayeska* – Great Land to the West – but this is disputed. One Aleutian island is called Unalaska, its name said to derive from the Aleut *Agunalaksh* which refers to the way in which the sea crashed against the local land, appearing 'to break its back' as, driven by the fast tidal races between the islands, it barged past.

Before considering American Alaska, it is worth noting that there is an ambiguity in the purchase as set down in the Treaty between Russia and the United States. Article I defines what exactly the Americans were getting in terms of the border between Alaska and Russia. (We shall return to the border in the central Bering Sea below, but here we consider the case of the Aleutian and Commander islands.) It states that the border will 'pass midway between the island of Attou and the Copper islands of the Kommandorski couplet or group in the North Pacific to the meridian of one hundred and ninety-three degrees west longitude, so as to include in the territory conveyed the whole of the Aleutian islands east of that meridian.' But Copper Island, and the associated Sea Lion and Sea Otter Rocks, lie *east* of 193°W. While the US government recognizes Russian sovereignty of Copper Island, realizing that an error was made at a time when longitudes were still occasionally subject to uncertainty, there are still Americans who routinely complain that their government is not attempting to pursue a claim to the island.

After the purchase of Alaska, the US Congress, which had been reluctantly convinced of the benefits of the deal, lost interest and the state remained a lawless place for two decades, policed by a few hundred soldiers who rarely ventured outside the security of Sitka and a few other well-protected enclaves. Adventurers and naturalists such as William Healy Dall (after whom Dall's sheep and Dall's porpoise are named) and John Muir did visit, but chiefly the Americans that came sought their fortunes in furs and whales in a manner which was not dissimilar to that of the *promyshlenniki*, and with much the same disdain for the rights and interests of the native peoples. Some also looked for gold. There were minor finds – the Russians had discovered gold in small quantities on the Kenai Peninsula in the 1850s, the first significant American find being close to what is now the State capital of Juneau (named after the French-Canadian Joe Juneau, who discovered the gold with his American partner Richard Harris). However, it was the discovery at Nome in 1898 that led to the Gold Rush and the first major influx of non-native Americans to Alaska (though the initial strike was actually by a trio who became known as the 'Three Lucky Swedes', Jon Brynteson, Eric Lindblom and Jafet Lindeberg, at Anvil Creek near the Snake River). The miners and the entrepreneurs, who made fortunes by supplying and servicing the needs of the miners (and by finding ingenious, not always legal, ways of relieving them of their cash), made Nome a boom town. One of the more famous inhabitants was the legendary gambler and lawman Wyatt Earp.

Prior to the discovery of gold at Nome, the major Klondike strike of 1896 had brought an estimated 40,000 men to Dawson City, a town which appeared virtually overnight in Canada's Yukon Territory. The men arrived by ship at either Dyea or Skagway, then proceeded via the Chilkoot or White passes to reach the Yukon River, which was navigable to Dawson City. The Canadian Mounted Police ensured a level of order amongst the throng and, equally importantly, ensured that all prospectors had enough supplies with them to avoid the chaos that would have ensued had ill-clad, ill-supplied men arrived in thousands at a remote town with few supply outlets. The ownership of the ports of Dyea and

Skagway caused a conflict between British Canada and America which, when resolved, finalized the border between the two countries.

The conflict turned on the definition of 'ten marine leagues' in the British-Russian agreement of 1825. More precisely it turned on exactly where the ten marine leagues should be measured from. Should some baseline be chosen or should it be from the head of each individual fjord. When the area was uninhabited and rarely visited, it hardly mattered. Now, with money to be made from those arriving at Dyea and Skagway it mattered a great deal. The Canadian view was that the two ports lay in Canada as it was ridiculous to measure from the furthest inland point of every fjord. The American view, not surprisingly, was that the furthest point was the rightful point. After much haggling, the British government, against the wishes of the Canadian government, accepted a compromise line between the two claims. The compromise, agreed in 1903, meant Dyea and Skagway were in Alaska, which also meant that a significant fraction of the wealth of the Klondike gold rush also went south to America rather than east to Canada. The Canadians were infuriated, but historians believe that overall the compromise was beneficial to all sides, particularly in promoting cordial relations between Britain and the United States and, when Canadian anger had subsided, between Canada and the United States. Though the 1903 agreement seemed to have established the border, one contentious area remained, and still remains. It is an area of sea lying south of Alaska's Alexander Archipelago and Prince of Wales Island, and north of the Canadian Queen Charlotte Islands. Known by the prosaic name of the Dixon Entrance, the area is a disputed fishing ground occasionally brought into sharp focus by the activities of US or Canadian fishing vessels deemed to be operating on the wrong side of the border, as seen by vessels of the other nation.

The northern border between the United States (i.e. Alaska) and Canada has also been contentious. As the border runs along the 141°W longitude line Canada considers that it should continue to do so once the Arctic Ocean has been reached, and that forms the basis of the 'sector approach' claim to the far north it set down in 1907 (see page 87). Given the proximity of the Prudhoe Bay oilfields it is no surprise that the United States disagrees, preferring a line at right angles to the coast at the border, the effect of which would be to move the maritime border further east (and so place more of the potential off-shore oilfields on the US side of the border). As with the Dixon Entrance, the border dispute is unresolved.

One further boundary was to cause problems, but not for almost a century. The treaty of 1867, which transferred ownership of Alaska from Russia to the

United States, included a border in the Bering Sea, but the actual form of the line is not mentioned and maps, from either side, are no longer available. There were no problems over the border line until 1977 when the USA and the USSR implemented 200-nautical-mile exclusion zones in the sea. The position of the Aleutian and Commander islands meant that for large section of the central Sea the exact position of the treaty line was now contentious. The problem was how the original line had been defined – had it been a straight line on a Mercator projection (a loxodrom line) or a straight line on the Earth's surface, the latter giving a curved line on a Mercator map (an orthodrom line)? Negotiations took place between the two countries and in 1990 the Baker-Shevardnadze Agreement (negotiated between US Secretary of State Baker and the USSR Foreign Secretary Eduard Shevardnadze – later, the president of independent Georgia) was concluded. Although a compromise between the two lines was negotiated, the agreed line favoured the US as it lay west of the mid-point line between the mainlands which is usually the starting point of negotiations between countries. Consequently the US ratified the Agreement immediately (though against the wishes of the Alaskan State government which believed that Alaskan interests were 'in jeopardy' if the Agreement was signed). The USSR did not ratify the Agreement, and in 1991 after its collapse, Russia claimed that the Agreement was null. The argument was that the Russia of 1991, newly emerged from the defunct USSR, had not been party to the negotiations and that as far as Russia was concerned – and it was Russia which bordered the Bering Sea – Shevardnadze had exceeded his authority in concluding it and Mikhail Gorbachev had acted against Russian interests in signing it as he was more interested in good relations with the US than in preserving Russian rights to the sea's resources, particularly in terms of fishing and mineral exploitation. The Russian Parliament therefore declined to ratify the Agreement. Meetings to obtain a balanced solution have continued since 1991, not helped by the activities of Russian fishing vessels, which have persistently entered waters which the US sees as its own (usually in an effort to avoid paying Russian export taxes by taking fish directly to markets in Anchorage or Vancouver), refusing US Coastguard efforts at inspection and, on one occasion, surrounding a Coastguard ship which was attempting to seize a vessel. The activities of American ships, which are caught fishing within Russian waters close to the Commander Islands, have not helped either.

Following the Nome Gold Rush, road and rail infrastructures gradually developed in Alaska. Settlers arrived attracted by the limited version of the Homestead Act applied to the territory; fortune-seekers were attracted by the possibility of gold or to mine existing sites (or new ones such as the copper discovered near the Chitina River); and tourists were brought by tales of glaciers, Mount McKinley, wilderness and wildlife. Soon newcomers outnumbered native Alaskans. In general the native groups fared better under the Americans than they had under the Russians, particularly after law and order (initially a 'frontier' form of each) was established. For one particular group, the Aleuts, this was, however, not the case. The pursuit of fur seals led to further abuses of the native population – a situation which was not helped by the US government's decision to make them 'wards' of the government rather than citizens of the United States, a status that was not corrected until 1971, twelve years after Alaska had joined the Union.

In June 1942 the Japanese occupied Attu and Kiska at the western end of the Aleutian chain after bombing Dutch Harbour on two successive days. Understandable American fears that mainland America was to be invaded by way of the Aleutian chain – a Japanese fleet had been spotted *en route* to Dutch Harbour at the time of the bombings, but had been intercepted and turned away – led to the evacuation of many Aleuts so that the islands could be fortified. The Aleuts were put into totally inadequate housing on the Alaska Peninsula where they endured a miserable few years. It is estimated that 10 per cent of the Aleut died of malnutrition and disease. Those that did return found that their houses had been looted or burned by occupying American troops who had also used churches and some houses for target practice. Claims for the damage dragged on through the courts for over forty years. Only in 1988, by which time there were only 400 survivors of the evacuation left, was a settlement agreed and compensation paid. Overall, it is highly probable that the Aleuts suffered more than any other Arctic native group from its contact with the 'civilized' world.

Alaska was allowed to send a non-voting delegate to the House of Representatives from 1906, but all attempts to acquire statehood were rejected: Alaska was just too remote, the population too small for the attempts to be taken seriously. The governor of the territory was a presidential appointee, which effectively made Alaska nothing more than a colony. Only in January 1959 did Alaska become the 49th State.

With the discovery of oil at Prudhoe Bay in 1968 Alaska experienced another 'gold' rush, with money and men descending on the state. The Prudhoe find required the construction of a 1,250-km (780-mile) oil pipeline from the Bay to Valdez, a remarkable feat of engineering which necessitated innovative solutions to the problem of building on permafrost. Oil revenue from the Prudhoe fields has helped to create an infrastructure and a standard of living in the State which is equal to that of the Lower 48 States (and better than many), though the downside, in environmental terms, was seen in 1989 when the *Exxon Valdez* ran aground and released 250,000 barrels of oil into Prince William Sound, a sensitive wildlife area (see pages 172–3).

The Prudhoe Bay oil discovery was the catalyst for the settlement of land claims by native Alaskans, though the beginning of such claims can be traced back to 1936, when the Haida and Tlingit Indians of south-eastern Alaska began a court case against the US government for compensation over its acquisition of the land that became the Tongass National Forest. The 1939–45 war, and procedural delays, meant that the case had not been heard before Alaska became a State in 1959. The case was heard shortly after, and the court found that the Indians had a legitimate case, but set the compensation for the land at a level based on the presumed value at the time of the purchase from Russia (50 cents per acre rather than the 2 cents that the whole state had cost). Not surprisingly, other native groups were dismayed, but little had happened before 1966 when the US Secretary of the Interior, Stuart Udall, took up the issue and placed a freeze on the transfer of all disputed lands to the State (or, by implication, third parties). As the most economical method of transferring oil from Prudhoe Bay was a trans-State pipeline, ice preventing ships from comfortably reaching the oilfields for much of the year, this decision put exploitation of the fields in jeopardy when they were discovered in 1968.

In the wake of the Prudhoe discovery native groups became increasingly vociferous, seeking not only restitution of their lands or significant compensation for their loss, but continuing payments from settlers or corporations who now occupied and exploited the lands, and made profits from them. Ironically, the native groups were supported not only by environmental groups, but also by the oil companies and a large number of officials from the Nixon administration. It was an unlikely mix, the thread binding them together being the need to get oil from Prudhoe Bay to Valdez without too much hindrance (oil men and government officials) and recognition that this need offered significant leverage in longer term objectives (native groups and environmentalists).

Ultimately, in 1971 the US Government passed the Alaska Native Claims Settlement Act (Public Law 92-203). The Act (its name usually shortened to ANCSA) rejected all claims to original ownership of land by the indigenous population in exchange for a sum of $962.5 million (a figure which seems to have been arrived at by no better method than the fact that $1 billion seemed a nice round figure, with a small subtraction for equally arbitrary reasons) and packages of land that amounted to 44 million acres (almost 180,000km^2). No future claims on land ownership would be allowed. All US citizens with at least one quarter native ancestry were entitled to a share of the benefits. Eligible folk did not receive benefits individually, but by becoming shareholders in one of thirteen created regional corporations (twelve within the State, the thirteenth for native Alaskans who now lived elsewhere). ANCSA has been amended several times since 1971, the intention of the amendments being to iron out some of the potential problems caused by the possibility of the Corporations failing, and to cover issues such as the sale of shares to non-native persons.

The native land holding granted by ANCSA sounds enormous, but actually equates to a little over 10 per cent of Alaska's land area. Whether this was adequate compensation depends on an individual's point of view. It can be claimed that before first contact the native groups 'owned' all of Alaska, but it can also be claimed that as it was mainly the coastal belt that was continuously occupied, inland Alaska being sparsely and only intermittently settled, the Act's percentage was fair.

Although it was the discovery of oil that finally brought ANCSA into being, the dispute over land had much earlier origins. After the US had purchased Alaska, Congress passed the Organic Act of 1884, which included the specific provision that 'Indians or other persons' should 'not be disturbed in the possession of any lands actually in their use or occupation'. In reality lip-service alone was paid to the provision, as gold miners, lumberjacks and the builders of houses, railways and factories soon showed. The government itself also laid claim to millions of acres of land for national forests and parks. As the quantity of land taken over by these and similar ventures grew, so did native opposition. But once ANCSA had been passed, the arguments on land usage were between those who wished to see large tracts of land becoming parks and wildlife reserves, and those who wished to maintain the right to develop Alaska. With the election of a pro-conservation President, Jimmy Carter, a bill was proposed which created over 100 million acres of new parks and reserves. This passed the House of Representatives, but met with sustained opposition in the Senate from those who considered it too anti-development. The Senate proposed a replacement bill which included substantial parks and reserves, but was much more understanding of those who wished to see Alaska's resources developed for the benefit of all. An impasse was reached, and rather than lose the bill completely Carter signed the Senate version into law in December 1980 (the Alaska National Interest Lands Conservation Act – ANILCA).

As might be expected, the passing of ANILCA has not led to the end of arguments over the development of Alaska. Today Alaska is at the forefront of environmental debate, its wilderness areas being among the most scenically beautiful and, from a wildlife point of view, important areas of the Arctic and sub-Arctic. National parks and reserves protect much that is most sensitive, but as the debate on drilling within the Arctic National Wildlife Reserve has shown, conservation status alone may not be sufficient protection in the face of determined exploiters.

Canada

The discovery of America and why the continent was given its name is, as all school children know, no secret, the names Columbus and Amerigo Vespucci being all that are required for the answer to an examination question. The reality is a little cloudier, perhaps involving fishermen from the Basque country and from Bristol, England, who are known to have fished cod on the Grand Banks before Columbus' voyage, and a Bristolian entrepreneur called Richard Amerycke. But what is clear is that the first incontrovertible landing on what is now Canadian soil was by the honorary Englishman John Cabot, who made landfall in Newfoundland in 1497.

Following Columbus's discovery, the Treaty of Tordesillas gave Spain and Portugal control of the Atlantic, and with it ocean trade with the Orient, something which made life difficult for the English, French and Dutch, who initially sought a different way by opening a North-East Passage along Russia's northern coast. Ice conditions prevented any realistic hope of a viable route and interest turned to the possibility of sailing around the northern tip of North America. The first to try was Englishman Sir Martin Frobisher who, in 1576, reached Baffin Island, making first European contact with the Inuit there (though Cabot had seen signs of native occupation on his voyage). Frobisher's meeting with the Inuit was peaceable at first, but then five of his men went missing and things became hostile. As already noted, Frobisher took a hostage, the first Inuit to visit Britain: during a damp, cold winter the man died of pneumonia.

Frobisher also made the first attempt at exploiting the riches of the Nearctic when he brought back a lump of black rock which was thought at first to be coal and then, to much greater excitement, gold-bearing rock. Or that's what Michael Lok, Frobisher's backer, said it was. Frobisher returned to Baffin Island in 1577 to gather more rock. He also brought back an Inuit family, man, woman and child. The man entertained Queen Elizabeth I by killing swans from his kayak before all three died of pneumonia. Despite the misgivings of some experts, the enthusiastic endorsement of the rock by an alchemist in Lok's pay succeeded in procuring Royal approval for a third voyage, but the tons of rock which were brought back proved to be valueless. It would not be the last time that the dream of Croesian riches from the Arctic would prove to be based more on hot selling than cold reality.

The idea of a North-West Passage did not collapse with the exposure of Lok's rock as fool's gold, and a constant stream of English ships headed west over the next few years, the names of Henry Hudson, Robert Bylot, Thomas Button and William Baffin being etched both on the history of the European discovery of Canada, and also on the map of the country. During his first voyage (in 1607) Hudson passed 80°N, a record northing that would not be surpassed for over 150 years, though it is a later voyage, when his crew mutinied, setting Hudson, his son and seven others afloat in an open boat which is the more famous (or infamous). The nine men were never seen again. Of the others, the most notable was William Baffin, whose 1616 voyage included the discovery of Lancaster, Jones and Smith sounds, a feat of seamanship so far ahead of its time that it was generations before his discoveries were validated.

Later, the crew of a ship commanded by the Dane Jens Munk became the first men to overwinter in Canada, an experience that only Munk and two others survived. Munk was followed a few years later by the expeditions of Englishmen Thomas James and Luke Foxe (whose names are also found on Canada's map), notable for the light they cast on the way in which seventeenth-century England viewed its position in the world. Each man was given a letter from King Charles I which he was to deliver to the Emperor of China. The letter was in English as it was assumed that the Emperor, being a king and therefore a cultured man, would inevitably speak the language. The voyages of James and Foxe were the last to be made in search of the Passage for some years. In England the realization that any route discovered was likely to be of limited value because of the vagaries of seasonal ice, together with the upheaval of the Civil War, made enthusiasm scarce, while French discoveries in mainland Canada suggested that the country itself was more valuable than a route around it. Beginning with the journeys of Jacques Cartier in the first third of the sixteenth century, the French had been exploring the St Lawrence River and the land which would become Québec.

The original intention of the French Crown was to settle the new land with farmers, but the incomers soon realized that Canada was rich in fur-bearing animals and had begun to trap them, offering Europe an alternative source of furs to Russia and the Moscovy Company. Chief among the animals trapped was the beaver, whose coat was not only thick but waterproof, making it ideal for hats and for northern European winters. Travelling down the St Lawrence River the French discovered the Huron Indians. History records that the names 'Canada' and 'Québec' derive from the Huron language: *kanata* is Huron for village, *kebek* for a river narrowing. The Huron were slash-and-burn agriculturalists, hunting being limited and usually only carried out in winter, but nevertheless were very willing to hunt beaver and other fur-bearing animals in exchange for French trade goods, particularly iron. The Huron were devastated by European diseases, just as all native peoples elsewhere were.

The French and the English also tolerated slavery among the native groups and, by preferentially arming one group, can actually be said to have aided it. It is also the case that the French discovered themselves to be in the midst of a continuous war between the Huron and the Iroquois to the north. It was therefore in their interests to maintain good relations with the Huron, their neighbours, and to support them during the frequent incursions of their enemies. French involvement in the conflict earned them the undying hostility of the Iroquois, which was later to become significant. It also meant that the Huron lost their language and became French speakers. Overall, disease and maltreatment reduced the Huron population by almost 75 per cent before conditions for the native peoples improved.

In England, after the Restoration, King Charles II became aware of the value of French fur imports from Canada (ironically from a pair of Frenchmen looking to make money by competing with their countrymen – see pages 112–13). After sending an expedition to Hudson Bay to confirm the fur richness of the mainland, he decided to organize his own trading empire. Under his nephew, Prince Rupert, the exiled King of Bohemia, a group of merchants was formed into the Governor and Company of Adventurers of England Trading into Hudson's Bay, a title soon shortened to the Hudson's Bay Company. The Company was granted 'sole trade . . . of all those Seas Streightes Bayes Rivers Lakes Creekes and Soundes . . . that lye within the entrance of . . . Hudson's Streightes'. The Company established a series of forts, each in the charge of a 'factor' (manager or governor) around the southern shores of Hudson Bay, the factor being charged with the task of trading furs with the local Indians. The Indians did most, usually all, of the trapping, and Company employees rarely ventured out to explore the local area, despite the Company also being charged with mapping northern Canada, one aim of this being the discovery of the elusive North-West Passage.

During the rivalry – albeit at a distance – between the Hudson's Bay Company and the French traders, the Iroquois sided with the British and increased their attacks on the Huron. Although the latter were as numerous as the Iroquois, they were no match for them in battle and by the time of the Canadian battles that accompanied the European war between Britain, Spain and France the

Huron had either been absorbed into the victorious Iroquois or had fled their homeland. The defeat of the French in Canada, together with the Atlantic blockade, gave Britain control of Canada and its fur trade, though Québec has remained resolutely French. Thereafter the history of Canada is, in many ways, the history of the Hudson's Bay Company, as its employees were responsible for much of its mapping, particularly in the Arctic. To such an extent is this true that the Company initials, HBC, were claimed to stand for 'Here Before Canada'. Its senior officials behaved, at times, in a fashion which mirrored that of the British Crown, not surprisingly because the Company was actually given jurisdiction over the country. The native groups which traded with the British were as susceptible to European diseases as any other, but in terms of the actual dealings between the two, the Indians fared much better than native Siberians did, at least until the Company began to trade alcohol for furs – drink being much cheaper and taking up much less room on supply ships than the weapons, iron goods, etc. which the Indians initially preferred.

One Company official, Samuel Hearne, followed the Coppermine River to the Arctic Ocean, while Alexander Mackenzie (of the North-West Fur Trading Company, initially a rival to the HBC before the two merged) followed the river which now bears his name to the Arctic Ocean. Mackenzie actually named the river Disappointment because he had hoped to reach the Pacific; but a few years later on a second expedition he did, becoming the first European to gaze out across the ocean from the American shore, and the first to have crossed the continent. During these and subsequent explorations, relations with the Inuit were largely cordial, as the Inuit were keen to trade for European goods. There were occasional fights, with each side taking advantage of a situation to obtain goods without the bother of trading for them, but these were not frequent. The Inuit were thinly spread across a vast area, so it was not easy for them to amass a sufficient number of men to seriously threaten a British fort.

With mainland Canada becoming better known, the British now re-entered the quest for a North-West Passage. As already noted, the hope of finding a Passage that was a worthwhile trading route had long since evaporated. But with the Napoleonic Wars ended there was a need to keep the Royal Navy occupied. Concerns were also growing about Russian objectives in the Southern Ocean where they had recently dispatched an expedition. If the Russians were to challenge British naval supremacy, a back door route to Siberia via northern Canada would be extremely useful. And so, over a period of thirty years a number of naval expeditions headed north-west from Britain, making household (and Canadian map) names of men such as Frederick Beechey, John Franklin, William

Parry, John Ross, James Clark Ross and Edward Sabine among others. Together with HBC employees such as Peter Dease, Thomas Simpson and John Rae, these expeditions discovered and mapped much of Canada's vast Arctic archipelago. But despite these successes, the main aim of the exercise, the discovery of a North-West Passage, remained elusive. The complex geography and the different distribution and thickness of sea ice, year on year, defeated all attempts. In part the defeats were due to the British use of large naval vessels, which were not well suited to the tight manoeuvering sea ice often required.

Finally the British decided to give up, but then, spurred as much by a sense of the national humiliation as by any remaining desire to find the Passage, they decided to try once more. They chose John Franklin, veteran of two successful Canadian land expeditions (in which the Navy had sent simultaneous land and sea teams) as leader. He was given two ships and a combined crew of 128. On 19 May 1845, as the ships were preparing to depart the River Thames a dove flew down and perched on the mast of one of the ships; the entire expedition complement cheered this happy omen. On 26 July Franklin's ships encountered a whaler near the entrance to Lancaster Sound. That was last time either the ships or any of the crews were ever seen alive by European eyes.

The search for Franklin or evidence of his fate occupied the next ten years, involving not only Royal Navy ships, but American expeditions and, eventually, one organized by Franklin's grieving widow when the British Admiralty declined her request for one more attempt. These expeditions finally found the North-West Passage (though they did not complete a legitimate transit) and mapped much of remaining archipelago. However, news of Franklin's fate came from none of these (though it was confirmed by Lady Franklin's private expedition). It came from John Rae, an HBC employee, and it said much about the British attitude to the Inuit. Rae brought back to Britain relics of the Franklin expedition he had obtained from the Inuit, along with their tale of men dragging heavy ships' boats along the west coast of King William Island and, ultimately, resorting to cannibalism – human flesh had been found in cooking pots – when their food supplies ran out. The British were appalled. *The Times* thundered that no one could take the testimony seriously as the Inuit 'like all savages are liars'. The author Charles Dickens was equally outraged, claiming the story was bound to be false as it was based on the word of 'the savage' and 'we believe every savage to be in his heart covetous, treacherous and cruel'. Dickens went as far as hinting that it was more likely that the Inuit had murdered Franklin's men, and that if there was indeed human flesh to be found it was an offering by the Inuit 'to their barbarous, wide-mouthed, goggle-eyed gods.'

Ultimately, Lady Franklin's private expedition supported (apart from the suggestion of cannibalism) the Inuit testimony, but that did not help John Rae. Despite a strong claim to having been first to identify a North-West Passage, and deserving a knighthood for his discoveries in northern Canada, Rae was shunned for having sided with the Inuit and died a humiliated, embittered man.

The North-West Passage or, rather, one of several passages, was eventually navigated by Roald Amundsen in 1903–5. Amundsen was probably the greatest of all polar explorers. (He was first to the South Pole, and arguably the first to see the North Pole, as his airship expedition definitely passed over it, whereas the claims to have reached the Pole by Cook and Peary are suspect. He also completed the third transit of the North-East Passage.) For the North-West Passage, he used a small ship with a crew of seven, including himself. The advantages of a ship that could cope with shallow seas, manoeuvre quickly, and with a crew small enough to live successfully off the land, showed the error of the Royal Navy's attempts. In their defence, the Navy were using what they had available to them, though their inability to learn from their mistakes – which Amundsen most certainly did – is an indictment.

The British had attempted to reach the North Pole as early as 1773 when two ships under Constantine Phipps sailed north past Svalbard, a trip most notable for having almost ended the life of a fourteen-year-old midshipman called Horatio Nelson, who attacked a polar bear (because he wished to take its skin back for his father) and discovered that the bear was more than capable of looking after itself. Only a shot fired from the ship persuaded the bear to run off. Nelson was lucky; polar bears live in an environment where sharp cracking sounds are, from ice breaking, not unusual, and single gun shots are not always a useful safety device.

The British tried again in 1827 William Parry taking a ship to Svalbard and then trying to advance northwards by hauling ships' boats across the ice. As others were to find later, sea ice drifts, and Parry found that for every one kilometre of forward travel his teams had to walk three. After the failure to find the North-West Passage, and the catastrophe of Franklin's last expedition, the British, somewhat surprisingly, made one last attempt to reach the Pole, sending George Nares and two ships to northern Ellesmere Island in 1875. It was the first of several attempts to reach the Pole using northern Ellesmere, or the nearby coast of north-west Greenland, as a starting point. The expeditions, chiefly of Robert Peary, mapped the northern coast of Ellesmere as a by-product of their attempts.

Following the failure of the Nares expedition, the British lost interest in Canada's Arctic, and in 1880 the government formally ceded their rights to the

Arctic islands discovered by its North-West Passage expeditions to Canada. It was hoped that Canada would take control of the area which had effectively become a 'no-man's land', one example being the use of Herschel Island as a base by American whalers, even though the island was Canadian territory according to the agreed boundary between Alaska and Canada. But Canada had more pressing problems, and did not immediately press its claims for ownership. At the time the decision seemed of little consequence, but it was to prove to be important.

In 1898 the Norwegian Otto Sverdrup, who had been captain of the *Fram* on Fridtjof Nansen's attempted drift to the North Pole, took the ship to northern Canada and during a four-year expedition mapped 250,000km² of the Arctic archipelago. Axel Heiberg and the Ringnes Islands were discovered and Ellesmere Island's west coast was explored. In the absence of any suggestion of ownership by Canada, Norway claimed the finds for itself. The United States, concerned that another nation might acquire lands close to North America, disputed the claim, but rather than supporting any Canadian rights it suggested that it might claim ownership for itself. Now, finally, Canada decided to assert itself. It claimed all lands northwards from its mainland coast to the Pole. To authenticate the claim it decided to buy Sverdrup's charts and maps and made an offer for them. Negotiations dragged on for years and were only finally settled in 1930 when Norway formally ceded its claim to the islands in exchange for $67,000, which was to be paid to Sverdrup. Otto Sverdrup was by then seventy-six years old. He had worked until he was seventy-two because the Norwegian government had declined, until then, to give him a pension despite his work in the Arctic. The money was finally paid to Sverdrup on 11 November 1930: he died fifteen days later, but had been grateful for the cash as it gave his family security. In 2004, in a joint venture between the Canadian, Greenlandic and Norwegian post offices, a series of stamps were produced to celebrate the 150th anniversary of Otto Sverdrup's birth.

The agreement between Norway and Canada is interesting as in international law the fact of discovering land does not, of itself, confer the right of ownership, greater weight being given to occupation and administration over a five-year period. Technically, therefore, it was probably sufficient for the Canadians to establish bases on Ellesmere and what are now called the Sverdrup Islands. But Arctic islands expose a flaw in the international position as their occupation and administration is no trivial exercise. Axel Heiberg is very mountainous, the Ringnes Islands very remote. Although it is likely that each had been visited by Inuit hunters, there was no settled population, or any realistic hope of one, in the twentieth century. In 1904 Canada had decided to underwrite an expedition to the (then unreached) North Pole and had bought the *Gauss*, the ship which had taken Erich von Drygalski's German expedition to Antarctica in 1901–3, and renamed it *Arctic*. The expedition was to be commanded by the Québecois Joseph-Elzéar Bernier. But formal claim to the entire Arctic archipelago was now seen as more pressing, and the Canadian government decided to visit the islands and make formal declarations. Bernier was therefore charged with making a series of journeys to the various islands, collecting tax revenues from whalers and traders, and ensuring the consistent five-year presence that international law demanded for a territorial claim.

The *Arctic* also carried scientists and geographers, who collected information on the islands and made better maps. Historically important documents and artefacts left by expeditions seeking the North-West Passage were retrieved, old cairns from that era were rebuilt, new ones added, and Canadian flags were raised. In 1907, in a speech made in the Canadian parliament, Senator Pascal Poirier stated that Canada should formally lay claim to all land between lines drawn from the western and eastern extremities of the mainland to the Pole. This was the first time that the 'sector' based idea of Arctic ownership had been annunciated and was followed up by the most important act of Bernier's journeys: on 1 July 1909 he erected and unveiled a bronze plaque on Melville Island which, after the required five years of presence, laid formal Canadian claim to all the lands of the Arctic archipelago, a total of 740,000km^2. The plaque states that, with the unveiling, Canada had taken possession of 'the whole Arctic Archipelago lying to the north of America from 60°W to 141°W up to the latitude of 90°N.' The Canadians formally claimed a sector of the Arctic in 1925. In response, the USSR made a similar claim the following year. However, it must be noted that the 'sector' approach to Arctic claims has not received the unreserved acceptance of other nations, and the validity of sector claims in international law is controversial, particularly when applied to areas of sea and/or ice. Also worthy of note is the fact that during the exchanges between Norway and Canada over the sovereignty of the Sverdrup Islands, Norway, while recognizing the Canadian claim, pointed out that this recognition was 'in no way based upon any sanction whatever of what is named the "sector principle".' Given that Norway would benefit considerably from international acceptance of the principle, it may now, perhaps, regret the statement.

By the end of Bernier's voyages the eastern Canadian Arctic had been thoroughly explored and mapped, but little was known of the western Arctic north of Melville Island or west and north-west of Banks and Prince Patrick islands. Therefore, in 1913 the government underwrote the Canadian Arctic Expedition of Vilhjalmur Stefansson. Stefansson was (and remains) a controversial figure in Arctic exploration, admired for his tenacity and survival abilities in the north, but castigated for his outlandish theories on the ancestry of some Inuit groups and the possibilities of farming the tundra. His expedition, aboard the *Karluk*, has already been mentioned in relation to the survivors of the ship's sinking making it to Russia's Wrangel Island, and Stefansson's attempt to settle Wrangel. The aim of the *Karluk* expedition was the search for new lands in the Beaufort Sea, as evidence from sea ice currents and drift rates suggested that land masses, perhaps even substantial masses, were present.

Karluk proved to be an old and tired ship which may have contributed to Stefansson's decision to abandon her and most of his expedition team. On 19 September 1913, with the ship stuck in the ice of the Beaufort Sea near Hershel Island, Stefansson announced he was going to cross the sea ice to the mainland in search of fresh meat. It was an odd decision as he had previously announced there was no game on the local coast, and the few men he took with him seemed more suited to a search for new lands than a hunting trip. Those left on the ship watched as Stefansson and his small team disappeared. They were not to return, leaving the *Karluk* and those on her to a long and terrible, and, for seven of them, lethal, ordeal.

The survivors from the *Karluk* were rescued from Wrangel Island in September 1914. The Great War had started in Europe, and Stefansson was soon forgotten. By 1918 the *Karluk* disaster had also been forgotten. On 11 November the Great War came to an end, and in the same month Stefansson and his team – it had been augmented by that from a supply ship that had been sent out earlier and which Stefansson had reached – arrived in Alaska. They had been on the ice for five years. Their survival caught the public imagination, the disaster of the *Karluk* being overlooked in the mood of celebration. But the results of the exploration had been disappointing. Stefansson had hoped to find substantial new lands, but had discovered only the relatively small islands of Borden, Brock and Mackenzie, north-east of Prince Patrick Island. They were the last large islands to be discovered: the search for Canada's Arctic archipelago had ended.

As elsewhere, after the traders and trappers, the missionaries moved in to convert the native groups, some missionaries having a dual purpose, offering both evangelism and trade. For the Inuit the setting up of trading posts was to alter their way of life. Once entirely nomadic, they now had to embrace a life which revolved around the trading post. For Inuit and the Indians to the south, European trade was also detrimental in another way. The introduction of the gun saw a change in native attitude towards Arctic animals. When hunting had been both a precarious and dangerous way of life, it was necessary for every bit of the prey to be utilized. But the gun meant that a caribou, for instance, could be killed simply because the hunter needed a few strands of sinew. Often only the delicacies, the tongue or liver, were taken. As protection against wolves, their principal enemy apart from man, muskoxen form a circle, young animals at its centre. The wolves are then faced by an array of heavy heads and horns. Unless the muskoxen can be persuaded to break the circle and run, the wolves go hungry. For a man with a gun, the muskox circle is a stationary target, and many animals, sometimes whole herds, were shot. Prey species diminished and starvation, always a threat for the Inuit, followed the days of plenty that the weapons first offered. This was one reason the Canadian governments of the early twentieth century took action to relieve the condition of the Inuit and other northern peoples. But there were other reasons as well. One was the need to provide welfare to Canadian 'citizens' once Canada had laid claim to the Arctic archipelago, while another was protection of its interest in the mineral rights of the north. The gold in the Yukon and, later, the oil find at Norman Wells suggested the north was mineral rich, and it could hardly be developed in the absence of some consideration for the native groups.

Initial Canadian efforts on behalf of the Indian groups were similar to those made by the Americans to their native folk. Treaties were signed and reservations set up, but in general those sections of the treaties which promised things in exchange for peace and cooperation, were ignored or proceeded so slowly as to be useless. One Indian chief, complaining to an official about the condition of his people said 'You can see how unhappy we are, how miserable and sick. When I made this treaty with your government I said that we should have a policeman and a doctor. You have sent nothing but missionaries.' Policemen and doctors cost money; missionaries are free. It is ironic that during the period until the outbreak of the 1939–45 war the Canadians (agents of a British Crown which considered itself to be the most enlightened nation on the planet) spent less, per head, on education, health and welfare than Alaska on its native people, and considerably less than Denmark spent in Greenland. In Russia, native groups had better education (even if it was largely overtly communist) and welfare (with clear provisos).

In view of the Indian chief's comment on the lack of a policeman, the one area in which Canadian government spending outstripped that of Alaska and Denmark was policing, largely to ensure that the fur trade continued without hitch and that no problems were encountered by prospectors. Applying Canadian law to Inuit communities brought problems because it was often at odds with the system of justice that had controlled Inuit affairs for millennia. One famous case involved the fur trader Robert Janes, who set himself up at Pond Inlet at the northern end of Baffin Bay. When the ship he was expecting to arrive to bring supplies and take away his furs did not arrive after three years, Janes became a desperate and violent man. The local Inuit, concerned for the welfare of their women when they had to leave on an extended hunting trip, did as their customs decreed. They killed Robert Janes. The man who struck the blow was arrested, tried, convicted and sentenced to life imprisonment in Winnipeg, a sentence he barely comprehended. He soon contracted tuberculosis in the alien environment of a southern prison; he was returned to Pond Inlet on compassionate grounds where, soon after, he died.

In an effort to aid the assimilation of the Inuit into the modern world – education, in Canada as elsewhere, being directed at children, adults having to fend for themselves – the Hudson's Bay Company prepared *The Eskimo Book of Knowledge* (published in 1931). Though well-intentioned, and stating much that was both true and relevant, the book was redolent of the paternalism that infused all dealings with the Inuit. Its introduction read:

> The Book of Knowledge is a token of friendship provided for you and for your family by the Governor of the Company. He is a man of great understanding and wisdom who decides the difficult problems of the Company and directs the traders in their duties.
>
> Being a good citizen, loyal to the King and to those who rule the British empire for the King, he wishes that you and your children, who are also citizens of the British Empire, should learn more of this Empire to which you belong, so that you may fully share our pride in the King who lives in the Mother Country far beyond the seas; and he wishes that you should also the share the King's pride and our pride in those parts of the British Empire called Canada and Labrador of which you inhabit the northern regions. Furthermore he wishes that by your good actions and by your mode of life you should add your share of honour to the British Empire.
>
> Being also a happy man rejoicing in his children and in the love of his family, he shares with you the cares and the joys of your family; and by this Book of Knowledge he will surely diminish the cares of your family and add to your joys, if you are wise enough to pay heed to his advice.

In the first part of this book the Company will tell you and your children about the British Empire and about Canada and Labrador and how you are entitled to the privilege of regarding our King as your King.

In the second part of the book the Company, which has consulted with many Traders and the most learned Doctors and the Men of God, will explain to you the change which has come to your mode of living and will show you by what means you will bring better health and therefore greater happiness to your children and to yourselves.

In the third part of this Book of Knowledge will be shown to you the means whereby you may gain greater possessions in trade for the benefit of your children and yourselves.

Let those of you who can read, recite the book to those who cannot read. In your camps discuss the book; talk of it in your igloos at night when your pipes are lit. It is a good book and a true book – this Book of Knowledge.

The whalers came to your country; the Men of God came and the Traders of the Company came. They altered the conditions of your lives. The bow and arrows which your fathers used, you have discarded for rifles: the kayak and umiak which your fathers used, many of you have discarded for the wooden boats with engines: the rich seal meat and the deer meat which were the life blood of all your people, some of you have discarded for White Man's flour.

When you first see a hunter very far away on the ice with his dogs, it is some time before you can tell for certain in which direction he is moving.

It was the same way with the officers of the Government and with the officers of the Company who could not tell at first whether your people derived good or evil from the use of the things which the White Men brought to your country.

In those days also White Men knew not of the things which are likely to happen when a people such as yourselves suddenly begins to use the things which White Men gradually learned to use over a great period of time.

Take heed to what is written here, all you men and women of the North. Your people have not derived good from the use which you have made of the White Men's things. The things which have been brought to you are good things in themselves, but you have misused

> some of these things, so that to-day you are a feebler people than in the old days when your fathers did not know the White Men. Your sons are less hardy, your wives bring forth fewer children. There is sickness among some of you.
>
> Here you shall learn how you have brought this weakness about.

That the book, regardless of its merits or otherwise, was produced by the Hudson's Bay Company is no surprise. In the period to 1939 almost every initiative undertaken among the Inuit was fronted by either the Company, the Royal Canadian Mounted Police, or missionary groups, direct government initiatives trailing in a poor fourth.

The reduction in animal populations led to the introduction of regulations on hunting and trapping, the start of an environmental movement. The starvation that resulted from the lack of prey meant the government had to provide food. A watching world would not be impressed if famine meant the Inuit died out. But the government usually made local traders responsible for distribution, making the Inuit dependent not only on the hand-outs, but on the southerners (the 'White Men') who provided them.

In 1953, the Canadian government decided to move Inuit families to new locations in the north. This is the only example of large-scale movement of indigenous people by force in the twentieth century. The ostensible reason for this was the lack of game in certain areas, but the move had much more to do with establishing a Canadian presence in the High Arctic. Families were taken from Inukjuak in northern Québec and from Pond Inlet on northern Baffin Island, to Grise Fjord on Ellesmere Island and to Resolute on Cornwallis Island. The families were told that houses would be ready for them and that local wildlife was abundant. They were told that anyone who wished to return after one year could do so. But regardless of these 'sweeteners' the moves were compulsory (and forced). On arrival the families discovered that there were no buildings, wildlife was not abundant nor was it of species with which they were familiar, and the return after one year option had been rescinded (as it would damage the sovereignty claims which were at the heart of the decision). Grise Fjord was so far north – it is the most northerly settlement in Canada – it had weeks of total darkness in winter and weeks of perpetual summer sun, something with which the Québec Inuit were unfamiliar: the Inuit soon gave the settlement their own name – Aujuittuq, the place that never thaws. Forced to stay, the Inuit adapted to the local wildlife and have turned the two settlements into communities offering High Arctic tourism. The villages work, though it can be reasonably claimed that this is in spite of, rather than as a result of, government efforts. In the 1980s the relocated Inuit and their descendants initiated a lawsuit against the Canadian Government stating that 'there is overwhelming evidence to suggest that the central, if not the sole, reasons, for the relocation of Inuit to the High Arctic was the desire by Canada to assert its sovereignty over the Arctic Islands and surrounding area'. In the Soviet press this incident was named as 'Human shield in the Arctic'. In 1993 the government finally held a series of hearings into the enforced move, largely at the insistence of the Royal Commission on Aboriginal Peoples (RCAP) which had been set up in 1991. After the hearings the RCAP declared that the moves had been 'one of the worst human rights violations in the history of Canada'. The government agreed to pay $C10 million to the survivors of the moves and their families, but has, to date, declined to apologize.

Although the Grise Fjord and Resolute moves are very poor examples of government initiatives in the north, there were others which were more successful and, particularly towards the end of the twentieth century, schemes became well-founded rather than well-meaning. (It is, frankly, impossible to discuss the early schemes without mention of some which, at the very least, offered amusement to the Inuit: the introduction of chickens to one settlement, all of which were eaten by sledge dogs; the attempt to interbreed Tibetan yaks with hardy cattle species to produce an animal that could be herded on the tundra; and particularly the Inuk – Inuk is the singular form of Inuit – paid a sizeable sum to trail around an area where muskoxen had been introduced gathering wool snagged on low shrubs in order to develop a knitting industry: he failed to find enough to produce one sock.)

By the 1960s the Inuit had been granted voting rights. Suffrage rapidly resulted in Inuit demands for both compensation for land (particularly after the Alaskan Native Claims Settlement Act) and autonomy within their areas. In 1973 the government stated its readiness to discuss land claims. This announcement was used against the government when the local Cree Indians and Inuit objected in court to the James Bay hydroelectric project, which involved the construction of a number of dams on the La Grande River on the bay's eastern shore, and the alteration of local watersheds to provide enhanced river flow. An injunction was granted on the grounds that the government's willingness to discuss land claims made the project, which had not sought the agreement (or even involvement) of native groups, illegal. Although the injunction was soon overturned, the action led to the signing of the James Bay and Northern Québec, and the North-eastern Québec agreements, signed in 1975 and 1978 respectively. Each agreement included the payment of millions of dollars to native corporations.

The James Bay project, which resulted in the flooding of more than 10,000km² of local forest, and with the watershed alteration involved an area of over 350,000km² in total (almost 20 per cent of Québec's land area) went ahead. The flooding resulted in the release of large amounts of methyl mercury into the water, which poisoned the fish that were a mainstay of local communities. Even now, concentrations of mercury in ptarmigan and caribou in this area are the highest in any Arctic region. Native groups had also asked for an insurance bond against subsequent flash flooding, which Hydro-Québec refused on the grounds that water release would be controlled. In 1984 an uncontrolled release, coincident with heavy rain, led to 10,000 caribou drowning. Compensation and clean-up cost the company more than the insurance they had declined.

The Québec agreements were followed in 1984 by the Inuvialuit Final Agreement, which covered the native peoples of the Mackenzie delta and, most significantly, in 1993 by the Nunavut Land Claim Agreement which transferred a vast area of land and substantial funds to an Inuit corporation. This was followed in 1999 by the creation of Nunavut, an essentially self-governing region covering more than 2,000,000km² and constituting about 20 per cent of Canada's land area. Most recently (2005), a Labrador Inuit Land Claim Agreement, covering Labrador and Newfoundland, was signed, with similar provisions to those of other Inuit agreements. Finally, in 2007 the Nunavik Inuit Land Claim Agreement, covering Ungava in northern Québec, was agreed in principle: it was finally implemented in 2008.

For the visitor, Arctic Canada offers scenery as diverse as the Mackenzie delta, where the difference between land and water is often blurred, to the massive granite peaks of Baffin Island's Auyuittuq National Park. Axel Heiberg is mountainous, but the predominant landscapes of both the Arctic islands and the northern mainland are polar desert and tundra. Canada also lays claim to the one of the most impressive wildlife centres in the entire Arctic. At various times of year, Churchill, on the south-eastern shore of Hudson Bay, proclaims itself the Polar Bear Capital of the World and the Beluga Capital of the World, and during the spring it is an outstanding place to see Arctic bird species.

Greenland

The first landing of Norse adventurers in Greenland was in AD 982 when Eirik the Red arrived, having been banished from Iceland for the murder of a neighbour. After three years of exile Eirik returned with tales of the lush land he had discovered to the west. It is, of course, well-known that Iceland is mostly green, while Greenland is mostly ice, and equally often reported that Eirik's description of the new land he had found was as much propaganda as fact. But it is also true that Greenland's coastal belt, particularly in the south-west, is indeed fertile and vegetated. It is also the case that at the time of Eirik's arrival the Arctic climate was warmer than the pre-global warming temperature of a few of decades ago. Indeed, it is possible that this warming aided not only the Norse settlement of Greenland, but the spread of the Thule people in the High Arctic.

With a fleet of thirty-five ships loaded with supplies and settlers, Eirik returned to Greenland in 986. Twenty-one of the ships turned back or were wrecked, but fourteen reached the safety of the fjords near Qaqortoq. More Icelanders arrived in 987, and two settlements were established: Østerbygd, or East Settlement, at Qaqortoq, and Vesterbygd, West Settlement, at Nuuk. The ruins of Eirik's own settlement of Brattahlíð, close to the shore of Eiriksfjord across from Narsarsuaq, can still be seen. At its most populous the Norse settlements numbered about 4,000 people, about two-thirds of them in the East settlement. The settlers were farmers, rearing cattle, sheep and goats. After the establishment of Christianity sixteen churches, two monasteries and a cathedral were built. Arguably the best remains of a church are at Hvalsø, near Qaqortoq.

From Greenland the Norse made journeys west. Their sagas report the discovery of Vinland, though exactly where Vinland was still taxes historians. Certainly Norse travellers reached Newfoundland, where the remains of a winter camp have been excavated at L'Anse aux Meadows, the only indisputably Norse site so far discovered. In the winter of 1002–3 Snorri Thorfinnsson was born in a Vinland winter camp (perhaps L'Anse aux Meadows), the first non-native American to be born in the Nearctic. Many authorities believe Norse voyagers may have explored the Labrador coast and, perhaps, southern Baffin and south-eastern Ellesmere islands. Norse artefacts have been identified on both Baffin and Ellesmere islands, but these could, of course, have been carried there by the Inuit. In Greenland Norse travellers definitely reached 73°N on the west coast, where three cairns have been discovered on the island of Kingigtorssuaq, north of Upernavik. In one cairn a runic-inscribed stone recorded that the cairns had been built by three named Norsemen in April, suggesting an overwintering close to the site. April of which year is not known, but it was probably in the late thirteenth or early fourteenth century.

The Norse met and probably traded with Greenlandic Inuit whom they called *skrœllinger*, a word that might derive from *skral*, small or weak, or from Karelia, a district of northern Finland/Russia whose native inhabitants were short, stocky and dark. A small horde of Norse material was found during excavations in 1978 on Skraelling Island (named by Sverdrup in 1898), off Ellesmere's east coast.

The items seem too haphazard a collection to have been the result of trade, and it is thought they are more likely to have been the result of a Norse shipwreck or, perhaps, from an enforced overwintering when a Norse ship was trapped in the ice. Evidence from both the Norse sagas and Inuit tales suggest that dealings between the two peoples were not always friendly, each telling of conflicts, usually on a small scale, but involving deaths on both sides.

From the Norse sagas it is known that the last Bishop of Greenland died in 1378, and that the last recorded ship left the island in 1410. Exactly why the Norse settlements died out is still debated, but there is general agreement that by the middle of the fifteenth century Norse occupation of Greenland had ceased. The most likely explanation for collapse is a change in the climate, a colder Greenland making agriculture untenable. It is probable that most settlers left, but it is possible that some stayed, perhaps waiting for a ship which never arrived and dying of starvation or being integrated into the Inuit community. Not until 1472 or 1473 was a concerted effort made to discover if any settlers survived. In one or other of those years (the exact date is disputed) King Christian I of Denmark sent an expedition. The expedition ships were attacked by Inuit in kayaks so only a limited search was made, but no evidence of any Norsemen was found.

Early British voyages in search of the North-West Passage made spasmodic contact with the Greenlandic Inuit, but no effort to explore the island was made until 1721 when Hans Egede, a young pastor from Bergen, arrived. Egede had persuaded the Danish king (who then controlled Norway and Iceland, as well as Denmark) that acquisition of Greenland might be beneficial in the longer term and that an expedition to search for evidence of Norse settlers would reinforce Danish claims to the island. The king was convinced and Egede set off. He stayed in Greenland for almost fifteen years, but his stay had mixed consequences for the Inuit. One child who visited Copenhagen returned carrying smallpox, and the epidemic which followed killed 25 per cent of the population. Egede worked tirelessly during this period, caring for the sick and for orphaned children, but his attempts at converting the Inuit to Christianity, though much lauded at the time, are now viewed much less favourably. Details of his rewording of the Lord's Prayer and Christian phrases for a population which did not understand what bread was, and had never seen a sheep – 'Give us this day our daily meat' and 'Seal of God' for 'Lamb of God' – are a delight, but his view of the Inuit are much less charming. He thought them stupid and dirty, with a collection of disgusting habits. He thought their shamans were liars and swindlers, intent only on relieving poor, gullible people of their money, an opinion at odds with almost every other opinion. Shamanism appalled him, and those Greenlanders who persisted in following it, who wore 'pagan' amulets despite his directives or who grew tired of his endless Bible stories and threats of eternal damnation he beat with a rope. Worst of all he seemed incapable of understanding how the Inuit could be so (relatively) good and peaceful and how their society could work so well given their adherence to shamanism. He seems to have been genuinely appalled that the 'Christian' behaviour of this pagan community could exist, considering it shameful rather than admirable. That his enforced conversions were actually leading to the breakdown of social cohesion seems either to have escaped him or bothered him not at all.

Egede's plan for Greenland was that it should be settled by good Scandinavian Christians, but not only were people unwilling to come, but the Danish king decided that the isolation of this new section of his realm was ideal for the disposal of undesirables. To Egede's horror his settlement of Godthåb ('Good Hope', now called Nuuk) became a dumping ground for convicts, and the 'women of easy virtue' who accompanied them. The behaviour of the newcomers, who were not housed in a prison, was so dreadful that the Greenlanders, who held Egede responsible for their arrival, attempted to attack him. In an irony which may or may not have been lost on him, Egede needed protection from the wrath of his flock.

With agriculture almost impossible, supplies always short, and the inhabitants of Godthåb not well-suited to work, scurvy soon reduced the population, and the experiment of using Greenland as a disposal ground for convicts was not repeated. Instead, over time, a few traders and a number of Moravian missionaries arrived. The civilized life Egede had yearned for was eventually realised. Godthåb and other settlements grew up and Denmark installed a governor to ensure Greenland's smooth running. The island's climate did not allow wholesale immigration, so that, unlike Australia (for instance), where incomers soon overwhelmed and marginalized the indigenous population, the Greenlanders did not become a minority in their own land. But trading inevitably created winners and losers, and as the Greenlanders were no match for the immigrants in this new field of commerce, the growth of Danish settlements resulted in poverty and misery for many. The occasional arrival of British whaling ships hardly helped the situation, as the British sailors traded alcohol for ivory and skins. Such a ritual did these visits become that the Danish record notes theft of ships from the settlement harbours; Greenlanders were using them to visit the whalers and returned apparently more by good luck than good management as the 'crew' were inevitably drunk.

In 1818 the last Greenlanders to be contacted, the Polar Eskimo of the north-west coast, were visited by John Ross on the first British Navy expedition in search of the North-West Passage. Although the west coast had still to be mapped in its entirety, Ross' expedition essentially completed its exploration. The meeting of the British and the Polar Eskimo was also highly significant for other reasons. It is often said that the long isolation of the Inuit by the extensive tongue of the Inland Ice (Greenland's central ice sheet) which reaches Baffin Bay had caused them to be convinced that they were the only people on Earth, but this is now considered to be unlikely by most authorities. Contact, albeit limited, had been maintained over the centuries since immigration from Canada. The story has been given credence by the Inuit reaction to the British ships. The British had a Greenlandic interpreter on board, and he was able to explain that what the British assumed were shouts of greeting were actually cries of 'Go away'. The British went ashore and traded sledges and other examples of Inuit clothing and equipment (as well as measuring the Inuit). In Ross' book of his voyage there is an illustration which has much to say about this encounter of two civilizations (see page 30). On the right are the Inuit, clad in skins and with dog-drawn sledges. On the left are the British naval officers in dress uniform – tailed coats, buckled shoes and cocked hats. After the recent victory in the Napoleonic wars, Britain was then master of Europe, which could be said (and was by the British) to also mean master of the world. The Inuit were impressed with the finery of these imperialists. What the British, with their innate superiority were missing, was that the Inuit were masters of their environment. The British completely failed to understand what they were seeing and it was to cost them dearly in later expeditions to the Arctic and, later still, in their early expeditions to the Antarctic.

The east coast of Greenland is very different from the west. The shore is deeply indented, steep cliffs often falling directly into the sea so that there is little coastal plain. Arctic currents send ice along the shore making travel difficult and dangerous and keeping wildlife to a minimum. It is likely that the east coast was always inhabited, but the settlements were small, the populations always on the edge of extinction. By the time Europeans began to visit, Inuit only inhabited the area near Ammassalik and, it is now believed, the few people there would likely have died out had contact not been made. There are now several small settlements close to Ammassalik and Sermelik fjords. The only other native settlement on the east coast is Ittoqqortoormiit near the mouth of Scoresbysund. The village was created in 1925 (though there is good archaeological evidence for earlier Inuit occupation) when about seventy people were brought north from Ammassalik. The move was, in part, an answer to the problem of Ammassalik being hunted-out, but also an initial response to growing Norwegian occupation of the north-east (see below).

By the middle of the nineteenth century the east coast had been mapped, in a rudimentary fashion, as far north as Hold with Hope by Danish, British and German expeditions, and also by British whaling captains (the latter including the Scoresbys, father and son, after whom Scoresbysund, the world's longest and widest fjord, is named).

The latter half of the nineteenth century saw a number of land expeditions which were significant in terms of both Arctic exploration (Nansen's crossing of the Inland Ice) and the mapping of northern Greenland (the Peary expeditions). During his expeditions on the Inland Ice, and subsequently as he was making attempts on the North Pole, Peary succeeded in mapping much of the western north coast, though his 'observation' of an (illusory) channel separating the island of 'Peary Land' from the mainland was to be a factor in the deaths of three Danes from a 1906 expedition. The search for the bodies of their countryman was to lead to one of the great Arctic survival stories when Ejnar Mikkelsen and Iver Iversen survived for two years after being stranded on Greenland's north-east coast.

Despite the deaths and epic survivals, the Danes succeeded in completing the map of Greenland before the outbreak of the Great War, though further scientific expeditions filled in a few sections which lacked absolute clarity. With the return of peace, Norwegian fur trappers arrived in north-east Greenland. Finding worthwhile numbers of Arctic foxes, and a good supply of muskoxen for food, the trappers built huts and eventually decided, with official approval, to raise the Norwegian flag in June 1931 and lay claim to the area, calling it Erik Raudes Land (Eirik the Red's Land). The outraged Danes took their case for sovereignty over all of Greenland to the International Court of the League of Nations in The Hague in 1933. The Danish case was supported by the inhabitants of west Greenland and by the United States who had no desire to see Norway added to Canada and Denmark in their Arctic sphere of influence. The International Court ruled in favour of Denmark.

During the 1939–45 war Denmark was occupied by Germany in 1940 leaving the position of Greenland open to debate. Britain and Canada wished to occupy it, but this idea was vetoed by the, at the time neutral, United States. However, in 1941 the Danish ambassador to the United States signed an agreement, against the wishes of his government, which made the island a US protectorate. Soon after, the forerunner of the Sirius Patrol, a Danish army dog-sledge unit which annually travels from its base at Daneborg in the north-east around the north-east and north coasts, was begun, in order to ensure that Germany did not set up weather stations in remote areas. In 1943 and 1944 German units did set up bases, but these were destroyed after small-scale skirmishes involving ten or twenty men. Since 1945 the Sirius Patrol has continued to travel to north-east Greenland, in part as a way of reinforcing Danish sovereignty.

The war also saw US construction of seventeen military sites on Greenland, a militarization which continued post-war when the United States built several stations of the Distant Early Warning (DEW) system on the island (and at other sites across North America). Most of these sites have now been abandoned, but the Greenlanders complain of a legacy of spilled oil and toxic waste, and claim that there are higher cancer rates in villagers close to some of the sites. In 1951 the Americans built an air base at Thule, the construction of which required the forced resettlement of the Polar Eskimo from Pituffik and Dundas to Qannaq, 140km (87 miles) to the north. After ten years of operation the United States constructed a radar system, part of the Ballistic Missile Early Warning System, in defiance of the Danish government. The construction caused a political storm, but the furore was minor in comparison to that created on 21 January 1968, when a B52G bomber crashed on to the sea ice about 12km (7½ miles) west of the base. A clean-up operation followed, though it was not immediately announced that the plane had been carrying nuclear weapons in contravention of a US-Danish agreement on Greenland being nuclear-free. Further details of the crash and subsequent activities are given on pages 161–2. When details of the crash were finally made public the Greenlanders made a concerted effort to have the base closed. Nevertheless, the base lease was renewed in 2004.

The Danish position on nuclear weapons on the island had been clear for many years, though in the early 1960s they had tolerated the construction of a small nuclear reactor at Camp Century (part of the US Army Cold Regions Research and Engineering Laboratory) about 225km (140 miles) east of the Thule air base. The reactor, a type PM-2A, generated 1.5MW of electrical power as well as providing steam for camp heating. The research carried out at the camp was important, providing the first deep drills of the Inland Ice and so forming the basis of later studies of Earth's climatic changes. However, it is very likely that the camp was also part of 'Operation Iceworm' which considered the possibility of burying missile bases beneath Greenland's ice. Danish knowledge or involvement in the project is still unclear, as is whether missiles were ever installed. The reactor was apparently dismantled and removed in 1964.

In 1953 after a Commission which considered the opinions of the Greenlanders, an amendment to the Danish amendment made Greenland part of the Kingdom of Denmark, essentially a 'county' of Denmark. The United Nations were unhappy with the decision, having presided over a period throughout which colonies had become independent of their colonial masters rather than amalgamating with them, but, after an address by two Greenlanders supporting the move, gave their blessing. However, in 1973, when Denmark held a referendum on becoming a member of the European Economic Community (EEC), the forerunner of the European Union, Greenlanders voted overwhelmingly against, fearing, correctly, that entry would mean Greenland's rich shrimp-netting and fishing waters becoming a common resource for member countries. At much the same time Danes in Denmark were becoming concerned over the imperialist overtones inherent in the position of Greenland, despite the Greenlanders enjoying Danish citizenship.

Seizing the moment the Greenlanders sought home rule. This was granted in 1979 and in 1983 Greenland withdrew from the EEC. Home rule gives Kalaallit Nunaat (as the native people call their country – it means Land of the Kalaallit, the latter the name given to ethnic Greenlanders) a greater say in the running of their island than, for instance, the creation of Nunavut does for the Canadian Inuit. In November 2008 in a referendum on achieving greater autonomy, Greenlanders voted overwhelmingly in favour. The vote means that Greenlandic will become the official language and that Greenlanders will take control of the police and court system, as well as taking greater control over the island's natural resources, though still sharing revenues with Denmark. The share given to Denmark would be used to offset the current annual subsidy to Greenland (which currently stands at 3,500,000,000 Danish kroner, about £425 million). The Danish government supported both the referendum and the outcome. After formal passage through both the Greenlandic and Danish governments, the changes are likely to come into effect in mid-2009. The vote is seen by both Greenland and Denmark as another step on the road to full independence, tentatively suggested as likely when Greenland no longer requires the Danish subsidy, though some in both countries still wonder whether the island, with its population of around 57,000 will actually have the human resources to run such a vast land. Under the terms of the referendum, Denmark retains control over Greenlandic foreign and defence issues.

It was Denmark which, in 1997, signed an agreement with Iceland for the establishment of a line delimiting the two countries in the Denmark Strait, much of which is less than 400 nautical miles wide. As with the agreement between Norway and Iceland over the boundary line between Jan Mayen and Iceland, a box was created in the Denmark Strait within which the boundary line was defined by geodesic lines. There was no agreement on the sharing of mineral resources within the 'grey areas' defined by the box as there was in the Iceland/Jan Mayen Agreement. One interesting aspect of the Denmark Strait Agreement was the limited allowance granted to Iceland for Kolbeinsey, a small volcanic island within the Arctic Circle which lies to the north-north-west of the Icelandic island of Grimsey. (Ironically, Kolbeinsey is subject to such intensive wave erosion that the Icelanders feared it would disappear completely, with impact on the agreement with Greenland. The island measures only about 90m^2 whereas in 1616 it measured closer to 8½ acres, and ½ acre in 1985. The concerned Icelanders reinforced the island with concrete, in part to make a helicopter landing area. By 2006 the helipad had been eroded such that landings could not take place. Further stabilizing work is now planned.)

In the pre-home rule era Denmark had already negotiated an Agreement with Canada on the boundary between Greenland and Canada in Smith Sound, Nares Strait (Kane Basin) and the Robson Channel. In the latter, the distance between the two countries is only some 25km (15 miles). The Agreement, signed in 1973, also set up a well-defined box within which the boundary followed geodesic lines. While no specific agreement, in terms of an agreed percentage, was included on the sharing of any discovered resources, the Agreement did include the provision for the sharing of all relevant data, and the provision that, should resources worth exploiting be discovered, an amicable agreement would be reached between the countries. One issue not settled by the 1973 Agreement was the ownership of Hans Island in the Nares Strait. The island, uninhabited and only 1.3km^2 in area, straddles the agreed boundary between Canada and Denmark/Greenland and is claimed by both. Initially a minor issue, the ownership of the island became of greater significance after it was discovered that a Canadian oil company had been using the island as a base for the study of drilling platforms in the area, something which, apparently, came as news to the governments of both Canada and Denmark. Since then both countries have, at various times, sent expeditions to the island to raise flags and claim sovereignty. The present position is that the two countries are continuing with efforts to reach a long-term solution.

As well as maintaining responsibility for foreign and defence issues, Denmark also continues to subsidize the country as a whole. It is very likely that several settlements would not be viable without this injection of cash, a fact which tempers the call for full independence. A particular problem has been the rapid growth in the Greenlandic population. It is estimated that there were about 6,000 Greenlanders in 1800, this figure doubling by 1920. It doubled again by the 1990s and now stands at 57,000. Coupled with a retention of ancient hunting practices, this has led to a worrying destruction of island wildlife, particularly on the more heavily populated west coast. For the visitor, the Inuit of north-west Greenland are probably the closest to the original, pre-contact form. These descendants of Polar Eskimo live in a beautiful area – but then so do all Greenlanders, the island being scenically magnificent. Greenland also boasts the world's largest National Park, enclosing the whole of the north-east of the island, as well as the most northerly land so far discovered, the island of Oodaaq which lies 5km (3 miles) off the northern mainland at 83°40′34.8″N, 30°38′38.6″W. The island is about 30m (100ft) in diameter and 1m (3ft) high. Oodaaq is more a gravel/silt bank than a true island and it is possible that as sea ice melts further such banks will be discovered further north. However, as sea levels rise, such banks, and Oodaaq itself, will be inundated and cease to exist.

This shot clearly shows that an Arctic fox has severed its own paw to escape a leg-hold trap baited with a caribou head, a phenomenon commonly regarded as a myth. Northern Canada.

Chapter Three

Exploitation: Arctic Species as the Target

What is the right of a huntsman to the forest of a thousand miles over which he has accidentally ranged in quest of prey? . . . Shall the fields and vallies, which a beneficent God has formed to teem with the life of innumerable multitudes, be condemned to everlasting barrenness?

John Quincy Adams, 1802

For millennia the native hunter-gatherer communities of the Arctic lived in harmony with the wildlife on which they depended. But, as already hinted at in Chapter 2, the harmony was not entirely one of choice. The shamanism practised by all the peoples of the north encompassed both a physical and spiritual connection between the hunter and the hunted, one which required rituals of the hunter to ensure both the complicity of the hunted and the favour of the spirit world. But, in reality, it was the slowness of travel and the crude technology available to him which stopped the hunter from devastating the wildlife. Once guns had been introduced the killing rate accelerated in all parts of the Arctic, and the arrival of the snow scooter and the outboard motor allowed hunting to become easier and so accelerated the killing even further. Modern welfare, in particular the introduction of modern medicine, which reduced child mortality and led to longer average lifespans, resulted in a rise in the populations of native communities, which in turn led to an increased death toll of wildlife.

Prior to the arrival of Europeans, the populations of High Arctic species were, relatively speaking, large. But the animals were spread over a vast area of sea and ice. The Inuit were a nomadic people, moving across the land in search of these elusive animals. Consequently they left few permanent sites. The most often seen evidence of their passing are the tent rings of summer camps, or the rather more substantial remains of winter camps. These are usually littered with the bones of seals, walruses or whales. There is evidence of more permanent settlements, for example the magnificent ritual site on Yttagran Island, off the northern Chukotka coast, where Yuppiat Eskimo waited for the spring and autumn migrations of the bowhead whale, erecting skulls and ribs, and sacrificing whale meat in stone-built enclosures to beg for the return of the whales. Another settlement is the rather more haphazard 'boneyard' at Gambell on St Lawrence Island, a midden into which generations of Yuppiat threw the leftovers of slaughtered sea mammals.

The Gambell site is an impressive pile of bones, clearly representing the remains of thousands of sea mammals. But the midden is probably many hundreds of years old. When that is taken into consideration, the annual accumulation is seen to be much more modest. For comparison, any visitor need only walk over to the far side of Gambell to look at the village dump. While it is obviously the case that the debris there – old snow scooters and TV sets etc – is individually much larger, the fact is that the dump, detritus accumulated over the last fifty years, covers an area which is not so different. Modern consumerism creates mountains of rubbish in Arctic settlements just as it does elsewhere.

Modernity was also in evidence when the Europeans and Americans moved into the Arctic two centuries ago. Although it began as a search for trade routes, northern exploration rapidly became both a land grab and an exploitation of resources. In the twentieth century, and continuing into the twenty-first, the resources which were (and are) sought were (and are) the minerals required by industrial societies. In the first phase the resources were animals, though the scale of the exploitation was no less industrial.

Fur

The fur and hides of Arctic species had been the clothing of native peoples since their arrival in the north, and had been traded since first contact. But the trade accelerated during the medieval period, with Russia providing furs to Europe. So important was fur at exorcizing the chills of winter in those days before abundant energy sources and central heating, that it inevitably became a social emblem. The wearing of the very best furs was confined to royalty and the nobility: even today at times of pageant Britain's House of Lords is populated with nobles wearing capes of ermine, the black tail tips of the animal forming a pattern against the white of its winter fur. Commoners had to make do with lambswool. So important was fur, and of such standing, that it entered the stories of the age. In the original version of *Cinderella*, the heroine wears a slipper of *vair* (the old French word for fur) not *verre* (glass).

Of all the furs, sable was the most coveted as its coat was luxuriously warm, and, as already noted, the decline in the local population had been instrumental in the Russian crossing of the Urals and the annexing of Siberia.

The Fur Trade in Russia

Once in Siberia, the Russian fur trappers set to work with an enthusiasm which astonishes and appalls in equal measure when the numbers of animals slaughtered is set down. To the trappers, Siberia appeared limitless, its animal numbers likewise, but the combination of greed, a willingness to endure a hostile, occasionally lethal, climate, ruthlessness and ingenuity, wrought havoc on the populations. Hunting was carried out in winter when the animals grew thicker coats and, in the case of the stoat and the Arctic fox, changed to a white fur which was highly prized, and had to be completed before the spring moult which ruined the pelt. The *promyshlenniki* (see Chapter 2) as well as native peoples, who were often skilled archers, shot the animals, aiming at the head so that the body fur was undamaged. But the market rapidly pushed for the widespread usage of various traps, which increased capture rates. Leg traps also left the body fur intact, as did a variety of traps which dropped weights on to the skull and body, and nooses which let the prey suffocate.

In seventeenth-century Russia, most of population in the countryside were serfs (essentially slaves), owned by a landowner. The only people who enjoyed any form of freedom were the cossacks who lived on the fringes on the empire, and worked in agriculture as free men in exchange for having to protect the borders from enemy incursion. (The position was altered when Sweden attacked Russia, requiring Peter the Great to establish a regular army.) Cossacks were quickly involved in the fur trade, acting as the bodyguards (essentially hired muscle) for the *promyshlenniki* who frequently encountered problems with the native Siberians.

Many native groups beyond the Urals had habitually paid *yasak* (taxes) in goods, livestock or fur to Ghengis Khan and those who followed him. With Moscow now in control of Siberia, payment of *yasak* was transferred to the Tsar's coffers. The *yasak* demanded of native peoples when the Russians arrived in central Siberia at the beginning of the seventeenth century averaged 20–25 sable pelts annually for each man. Within fifty years, that number had been reduced to three, reflecting the reduction in the population. Local tribe leaders were responsible for the collection of *yasak*, the Tsar granting them the status of local dignitaries and sending them special regalia in exchange. In contrast to the majority of Russians west of the Urals, none of the Siberian peoples were ever subjected to serfdom. However, it must be borne in mind that the sable population was very large before the arrival of the trappers. The trappers who arrived at the Yenisey River had reduced the sable population up to that point to such an extent that trapping was barely worthwhile, discovered that sable were still very numerous. It must, though, also be noted that within thirty years of the Russians arriving there were so few sable left that trade in sable pelts ceased in the area.

From its inception, although sable was the preferred pelt, other furs – from fox, squirrel, ermine, bear, beaver, wolverine and wolf – could be substituted, the number of pelts being recalculated on the basis of worth against a sable. After collection, the *yasak* pelts were sent to Moscow (under military guard) where the Siberian Department sorted them into grades and decided on the price of onward sale. As well as *yasak*, the Siberian department also extracted a 10 per cent tax from the fur trappers, stationing officials who searched incoming trappers and traders for illicit furs concealed about their persons or in any of the myriad ways in which contraband furs could be smuggled. As the officials in the field could also be bribed by the merchants who employed the trappers, or could enter into illicit partnerships with the merchants, it is likely that a substantial fraction of the tax was avoided, particularly as most bribes paid to the officials were in furs, which then joined other clandestine exports. And it was not only the officials who required bribes, there being examples of members of the church who declined to take marriage or other services unless a bribe in furs was paid.

Uncertainty over the total number of furs collected, as opposed to the number passing through the Siberian Department, therefore makes it difficult to be sure exactly how many animals were being killed each year. By about 1600 around 20,000 official furs were reaching Moscow. As *yasak* made up about 75 per cent of the official number, tax on the merchants making up the rest, that would imply that the total number of animals killed was at least 65,000. When the whole of Siberia had been opened up it is estimated that perhaps ten times that number of animals were being killed annually: the trapping of sable alone probably exceeded 100,000 annually at its height. Accurate figures for the trade are rare, but one event, in 1595, does give an indication of its scale. In that year, Rudolf II, the Holy Roman Emperor, demanded soldiers from Russian Regent Boris Godunov to help with a crusade against the Turks. Godunov was unwilling to comply as the Turks were a major trading partner and close enough to represent a considerable threat from reprisal raids. So, to placate the Emperor, the Tsar sent a present of furs. The consignment consisted of 40,360 sable, 20,760 other martens (ermine, wolverine, etc.), 3,000 beaver, 337,235 squirrels and 1,000 wolves. When delivered, the furs filled twenty rooms of the Emperor's palace, with numerous wagons parked outside still loaded with the less-valuable squirrel furs. The exact fraction of the inventory of the Siberian Department is not stated, but it can be assumed that it was much less than an annual supply – even compensation to an Emperor would not go that far, particularly as fur was Russia's largest foreign trade and furs piled at the Department's Fur Treasury were the equivalent of the gold stocks of other nations.

As the supply of furs dwindled in occupied parts of Siberia, new hunting grounds became available to the east, as the Russians moved inexorably towards the Pacific. But eventually supply exceeded demand in Europe, especially after the Hudson's Bay Company had begun trapping in Canada. For a while Russia's economy, so closely linked to furs, looked precarious, but then both a new, and improved, source of supply, and a new customer were found to the east.

On Vitus Bering's second expedition the German naturalist Georg W. Steller became the first European to land on Alaska when, sensing history in the making, he leapt ashore on Kayak Island on the morning of 20 July 1741. During the voyage Steller made the first observation of several species which now bear his name, including Steller's jay; his name is also associated with Steller's (or northern) sea lion and Steller's sea eagle. He also made the first observations of the sea otter, and of Steller's sea cow, a large manatee which swam the waters off a pair of islands close to Kamchatka, the Commander Islands, named after Bering himself, the expedition's commander. By the time Bering reached the Commander Islands he was ill with scurvy, as were many of his men. On the largest island – now called Bering Island in his honour – he died, and he and those of his crew who also died were buried there. The deaths occurred despite the animal life the men found on the island – Arctic foxes, sea otters and the sea cows. The foxes and otters did not make good eating, but the sea cows did. What the men also found was that the pelt of the sea otter was the most luxurious fur they had ever seen. Spending almost its entire life in the sea, and not having an insulating layer of blubber, the sea otter had evolved fur with a hair density of up to 165,000/cm^2 and a total of *c.*800 million on the entire body. The expedition's ship, the *St Peter*, had been severely damaged and was not seaworthy. The surviving crew built another from salvaged timber and prepared to sail for Kamchatka. Recognizing their likely value the men took sea otter pelts with them, but during the voyage they were forced to jettison the pelts when the makeshift ship began to sink. But one man held on to his stock and when they reached the mainland he sold them for a price which shocked even him. The pelts, together with the opening of a new fur trade with China, brought a flood of trappers to the Commander Islands, the Aleutians and Alaska. Their relationships with the local population, and the introduction of foxes to the Aleutian Islands has been covered in Chapter 2. Here only the facts of the sea otter and fur seal trade are considered.

The first ship to sail to the Commander Islands, just two years after the return of the Bering Expedition, came back with 1,600 otter pelts, together with around 2,000 pelts of the northern fur seal (the fur of which is almost the equal of the sea otter) and 2,000 Arctic fox furs. Then, in the mid-1780s Grigoriy Shelikhov, a merchant from Rylsk (a town in the Kursk District in Russia, some 400km/250 miles south from Moscow) overwintered in the Commander Islands, taking 18,000 northern fur seal pelts, the sale of which gave him the capital to start the Russian-American Company. It is estimated that over the next twenty years the Company took 70,000 otters and 1,250,000 seals, as well as a huge number of foxes. In total that was about 70 per cent of the total Siberian production, an indication of the overhunting of the mainland: sea otter pelts accounted for 85 per cent of Russia's trade with China. But the killing came at a cost: in the years up to the end of the 1914–18 war it is estimated that between 500,000 and 1,000,000 sea otters were killed, reducing a population estimated at about 300,000 to just a few hundred; Steller's sea cow fared even worse, being hunted to extinction. The killing was not, however, entirely the responsibility of the Russians; after the transfer of the Aleutian Islands and the Pribilof Islands to the United Sates, American trappers hunted both the seals and the sea otters at a higher rate, and from 1871 onwards the Commander Islands were 'rented out' to a succession of private companies, initially to an American company (Hutchinson, Kohl & Company of San Francisco, 1871–91), but then to Russian companies (the Fur Seal Trade Company, 1891–1901, the Kamchatka Trading House, 1901–12, and the Vladivostok-based Churin Company 1912–16). By the 1920s it was feared the sea otter might be extinct, but a remnant population was discovered, and with the aid of full protection and relocation, the population has risen towards about 100,000 today. The hunting of sea otters is now forbidden.

If the numbers of sea otters killed seems astoundingly high, the take of northern fur seals involves even larger numbers. In 1786, Gavriil Pribilof spotted the islands that now bear his name; he sailed home with 40,000 northern fur seal pelts and 2,000 sea otter pelts, as well as over seven tons of walrus ivory. The islands' population of sea otters was rapidly reduced to zero, and by the 1830s even the fur seals, whose population had been very much higher at the time of discovery, were approaching extinction, perhaps as many as 2.5–3 million animals having been killed. In 1835 an edict forbade the killing of females and males on land, which limited the trade significantly, but when the Pribilofs became part of the United States unregulated hunting began again, almost 4 million seals being taken by 1911 when the four nations involved in the hunt – Japan, Russia, the UK (on behalf of itself and Canada) and the United States – agreed to manage the stock (see also Chapter 4). The agreement was not signed without rancour and resentment. Article I outlawed pelagic hunting which was resented by the Japanese and Canada/UK who chiefly operated in the open sea, while Russia and the United States were given control of the land-based hunt (not entirely unrealistic, given that the two countries had sovereignty over the land in question). Argument over the various Articles became so heated and irrational that one man – either British or Japanese, accounts vary – suggested, apparently without irony, that the ideal solution would be the hasty extinction of the species to avoid further conflict. To placate the British and Japanese, the US agreed to pay each $200,000 (worth about $4.5 million today). The agreement saved the fur seal and the population rose steadily, even with hunting quotas that varied over the years from 4,000 initially to 95,000 by the 1940s. Hunting of female seals was forbidden in 1968; this initially allowed the population to rise, though it fell again in the 1970s as a result, it is believed, of overfishing in the Bering Sea. Native Alaskans (the descendants of Aleuts forcibly resettled by the Russians as surrogate hunters) are allowed an annual quota of immature male seals.

As well as sea otters and fur seals, bears, beaver, foxes, lynx, mink, river otters, sea lions, wolverine and wolves were also taken on the mainland, a total of over 200,000 pelts from these species being exported in the years during which Aleksander Baranov was in control of Russian America. Assuming there was nothing particularly special about these years – other than Baranov's superior book-keeping – then over 600,000 mainland animals were successfully exported from Alaska to mainland Russia during the period of Russian control. Allowing for the assumed figure that 25 per cent of all vessels engaged in trade between mainland Russia and Alaska were lost at sea, the total number of animals killed is likely to have been in excess of 800,000.

Many men were also lost when ships went down, and it is worth noting that while the fur trappers were greedy, ruthless men and their treatment of the native Siberians and, especially, the Aleuts was atrocious, the attrition rate amongst the rank and file seamen and trappers was high. A fur trapping voyage might last four years (or even longer when the animal populations were declining, the ships staying out as long as it took to make a profit) during which time about 25 per cent of the crews died of disease or accident, as well as the 25 per cent who were lost in shipwrecks, so that only about 55 per cent of those who set out ever returned home. To the total number of animals killed by the Russians must be added another, but unknown, number killed by American adventurers who sailed Alaskan waters looking for sea otters and fur seals. Baranov frequently complained about the behaviour of these Americans, who, he considered, treated the native people appallingly. It is known that some of the Americans actively engaged in setting up a system of slavery among the Indian tribes of south-east Alaska, and also traded weapons with the native peoples and, as has already been noted in Chapter 2, while Russian treatment of the Aleuts was shocking in many respects, overall it was better than the treatment meted out by North American colonists.

Fur trapping remained both an important occupation and a regular source of foreign exchange for the Soviet government through most of its years in power. As the regime maintained a monopoly on currency exchange and the rouble rate was fixed arbitrarily by Stalin, the government was continuously short of the valuable exports required to maintain the trade balance: oil, precious metals and furs were the important commodities, furs often being called 'soft gold'. The invention of the snow mobile made the task of trapping easier, and the industry reached new heights in terms of killing rate. By 1980 it is estimated that there was one Arctic fox trap for every 10km^2 (4 square miles) of tundra, with corresponding numbers of traps for other animals, particularly sable. The fox kill often exceeded 100,000 animals per annum during the 1970s and 1980s, representing as much as 60 per cent of the total population. The fecundity of the fox could sustain such losses in years when the birth rate was high because of high prey density, but not in poor prey years, and not surprisingly the trade eventually caused a significant fall in fox numbers. Since man first trapped animals for their fur, the preferred method was the leg-hold trap as this did the least damage to the pelt. But it causes terrible suffering as the animal rarely dies quickly; as a consequence of this cruelty the European Union (EU) banned the import of fur caught using leg-hold traps in 1991, though the ban is considered ineffectual as there are loopholes, and it is impossible to police as examination of a pelt gives virtually no indication of the method of trapping.

Leg-hold traps also take a toll of wildlife other than the species for which they were set. Gyrfalcons and snowy owls, both symbols of the Arctic, were caught while attempting to take the bait or, occasionally, when they used the traps as convenient perches. One estimate, based on capture records, gave a figure of one gyrfalcon killed for every fifty traps, a mortality highly significant in terms of the overall number of falcons. The change of consumer attitude in the West towards fur and its methods of production caused a decline in Soviet and Russian exports, but fur is still used extensively in the country itself as a less expensive way of combating the frightening cold of the Russian winter. It is also the case that since 2003 the trade has revived somewhat, the price of a sable pelt increasing to $150–200 in 2008.

The Fur Trade in North America

The killing of fur-bearing animals for clothing was as important to northern native Americans as it was to the northern peoples of Eurasia, and it is known that as early as the late sixteenth century North American furs had been traded by Basque fishermen and others that ventured to the New World. But commercial

The Arctic National Wildlife Refuge (ANWR), which as been central to the debate on the drilling of sensitive Arctic areas.

The Mackenzie River, north-east Canada. The river's delta is Canada's premier source of natural gas.

Inuvik in Northwest Territories, Canada. The town's airport is a part of the Canadian Air Force's Forward Operating Base.

A muskox herd grazing in front of an oil-drilling complex, Prudhoe Bay, Alaksa.

Sacrificial enclosure of stone and bowhead at Whale Alley on Yttagran Island off Russia's Chukotka coast.

Bowhead whale carcass, St Lawrence Island, Alaska.

Skinned seal heads, south-east Greenland.

ABOVE Snow scooters and rifles have helped native Arctic dwellers to travel further and so reach, and kill, more animals.

LEFT Chukchi people erecting a reindeer skin *yarang*, Chukotka Peninsula, Russia.

BELOW Nenet people, Yamal Peninsula, Russia.

LEFT The Boneyard, St Lawrence Island, Alaska.

BELOW It is easy to persuade shipping companies to take large consumer items to the Arctic, but no one is interested in removing worn-out rubbish, so settlement dumps grow relentlessly.

ABOVE The old Blue Lagoon, Iceland. Iceland has an abundance of geothermal power.

LEFT The Smoking Hills, Franklin Bay, Northwest Territories, Canada. Here bituminous shale was ignited, probably by lightning, many thousand years ago, and still burns. Oil shales are now, controversially, being mined in Canada.

ABOVE *50 Let Pobedy,* Russia's newest nuclear ice-breaker, during construction in St Petersburg.

BELOW Nikel, the metal smelting town close to the Norwegian border on Russia's Kola Peninsula (photo: Timo Halonen).

Midnight sun, Lancaster Sound. In the background is Cornwallis Island. The Sound is the start (or end) of the North-West Passage.

exploitation of Nearctic species did not begin until the seventeenth century when the French began to exploit beaver. Beaver fur has the advantages of being waterproof and durable, which made it ideal for hats and the trimmings of 'outside' wear. Its usefulness in that regard had already reduced the population of European beavers to the point where their trapping had ceased to be economic, so the discovery of a new source was timely (unless, of course, you were an American beaver). From their settlements on the St Lawrence River, the French controlled the North American fur trade for about fifty years. Then, two dissatisfied French trappers gained an audience with King Charles II, the recently restored English monarch, convincing him that the fur trade in Canada was worth a fortune, one in which the two men could share. The two men have gone down in Canadian history, though this is, in part, a consequence of the English inability to spell their names correctly (or an indifference to getting them correct). One English spelling of Groseilliers was Gooseberries. As his companion's name was Radisson, Canadian schoolchildren now learn of the importance of Radishes and Gooseberries to the history of their nation.

The English sent a ship to check on the Frenchmen's story. They returned with the news that it was true, and the king, in a display of monarchical power that would have impressed his father, decided that England should take charge of Canada, dismissing with little consideration the claims of both the native peoples (who were ignored) and the French (who were, eventually, ousted from control by force, though they have maintained a foothold in Quebec). The king gathered together a group of interested merchants and under the governorship of his nephew Prince Rupert formed the Hudson's Bay Company.

The Company established a series of forts across Canada, each under the control of a 'factor' or manager who was charged with trading furs with the local native people. Initially trade was good as there was much the Europeans had that the Huron, Cree and other Indian groups wanted. Iron goods, cooking pots, knives and axes, were of prime importance, but so were trinkets – glass beads and so on – which could easily replace porcupine quills and other hard-to-work items which Indian women had used for centuries to decorate clothing. The history of the last two centuries in the Arctic includes many ironies. Not least of these is that when they first came, outsiders traded baubles for things of great value – furs, ivory and so on. Today many outsiders arrive as tourists and exchange things of great value – pounds and dollars – for goods manufactured solely for the tourist trade, items which amount to little more than baubles.

During the time of the HBC the Indians did almost all the trapping, the Company men staying within the confines of the fort, often a necessary precaution as factors were tasked with getting the very best deals they could, which created resentment among the Indians. Eventually, the Company was forced to supply guns and other more expensive items when the stock of beavers and other fur-bearing animals began to dwindle, but then realised that the supply of alcohol was cheaper. The Company's motto was *Pro pelle cutem*, which defies a true translation but was devised to mean something along the lines of 'a skin for its equivalent'. When the native hunters became a little more astute in their dealings with the Company, the factors were more inclined to suggest the motto meant 'we skin you before you skin us', though the native trappers preferred 'we risk our skins for your pelts'. Later still, when alcohol increasingly became the trading tool of choice, the motto was said to mean 'a skin for a skinful'.

The HBC traded chiefly on the mainland, but there was also trade with the Inuit, though this was less well organized and, consequently, more sporadic. Sealskin clothing and skins were traded and there was an increasing trade in Arctic foxes, particularly the white winter pelts, as animal numbers declined to the south. The result was, inevitably, a decline in fox numbers. The major impact of European exploitation of Arctic wildlife on the Inuit was, however, when the whalers arrived: this is dealt with below.

Because the Hudson's Bay Company trade was better controlled than its equivalents in Siberia, and was also less open to corruption, the take of fur-bearing animals is better quantified in North America (though that is not to say there was no illicit trade or smuggling – there was, but to a lesser extent). In the century that began in 1769 the Company exported to London almost 5 million beaver furs, together with 1.5 million mink, 1.25 million pelts of marten species, over 1 million lynx, almost 900,000 fox, almost 500,000 wolf, 288,000 bear and 275,000 badger. Though high, these figures are some five times lower than the estimated Siberian mainland take. If the sea otter and fur seal figures are also considered, then exports from Canada were perhaps ten times lower than those from Russia. However, the export figures do not tell the whole story. Once they had guns, the Canadian Indians behaved exactly as the other native peoples did, shooting animals just to extract the delicacies, the liver or tongue, or often just for the sport of killing. Those Company men who went outside the forts did much the same. It is impossible to gauge the numbers of animals involved, but over the long history of the Company it must have run into tens of thousands. The animals slaughtered for delicacies and trophies differed from those killed for their fur. The herds of caribou which roamed the forests and tundra of central and eastern Canada were destroyed, as were the moose. Sometimes a caribou would be shot just for its sinews, the remaining carcass being left for scavengers. Ultimately the total slaughter was such that animals became scarce and the Indians ran short of food. Starving Indians occasionally looted the Company forts and, in an ironic move, the Company was actually forced to ship food from Britain to Canada to feed native families, the decision mixing common humanity with the need to both protect and feed its employees. But despite the Company's best efforts, starvation of Indian groups became commonplace throughout the nineteenth century.

In the 1840s Prince Albert, Queen Victoria's consort, popularized the wearing of the silk top hat and almost overnight the demand for beaver among the upper echelons of British society dropped. It was probably just as well, as beaver numbers had gone into sharp decline. Nevertheless, beavers were still trapped, in reduced numbers, for many years until, finally, the reduction in population meant that the trapping was no longer economic, even in new areas where the animal had previously existed unharmed. In northern Quebec, for instance, the capture rate near Rupert House, one if HBC's trading posts, had been approximately 2,000 animals annually, but had begun to decline in the 1920s; in 1929 just four beaver were trapped.

The economic crash of the 1930s, the 1939–45 war and then the development of central heating reduced the need for furs in the West, while fur farming (for instance of mink) and the import of nutria (the fur of the coypu) reduced the need for wild capture. This and, as noted above, a change in attitude towards fur clothing in the west led to a decline in wild trapping. However, trapping continues in North America, the Alaskan trade being around 1,500 foxes annually (though much higher numbers are occasionally quoted), with higher numbers of animals being taken in Canada in a trade which is still claimed to employ 70,000 people, including 60,000 trappers. Another irony of the Arctic is that today the Canadian Fur Institute campaigns against the destruction of habitat of fur-bearing animals, on both aesthetic grounds (the loss of the animals) and economic grounds (the loss of income of its members).

The Fur Trade in Greenland

The trapping of Arctic foxes has never been as important in Greenland as the hunting of seals for the indigenous people, though the fox was the primary interest of the Norwegian trappers who built huts in the north-eastern part of the country in the inter-war years. The dispute between Denmark and Norway over the claims to the area (Eirik Raudes Land) has already been mentioned. There were not many trappers, nor were the foxes killed great in number, though those trappers who worked Jan Mayen rather than Greenland succeeded in driving the fox to extinction on the island. Jan Mayen would have been inhabited by foxes using sea ice bridges from Greenland and Svalbard; the retreat of the sea ice since the time of the trappers has meant that recolonization has not been possible. Given the current acceleration of sea ice retreat, the Arctic fox will never again be a resident of Jan Mayen.

On Greenland the trade in both seal and Arctic fox skins was controlled by the Danish Royal Trade Company (KGH) until the establishment of home rule. It was then controlled by Great Greenland, a company owned and operated by the Greenlandic home rule government, who were buying around 75,000 seal skins, but only a few hundred (at most) fox furs in the 1990s. By the end of the millennium, this number had fallen to barely 100 in total, due to a fall in demand rather than a decrease in animal populations, though the latter did have an effect. Then, during the early years of this century, driven mainly by demand from China, the numbers of skins purchased by the government rose again. By 2005 the number of seal skins exported annually had risen above 100,000. The rise has continued, but shows signs of having peaked at about 125,000 annually.

Whaling

Stone Age rock carvings of whales show that coastal peoples of the Neolithic age knew of them, though it is likely that their knowledge came from animals washed up or stranded on beaches. It is possible they hunted whales close to shore, perhaps using boats to drive schools of whales onto shallow beaches, a technique which was certainly in use in Norway in the ninth century, and is still used in the Faroe Islands today. Whale Alley on Yttagran Island dates from a later period, and shows that the Yuppiat were hunting bowheads many hundreds of years ago, though it is very probable that advances in technology, particularly the use of metal, was required before such huge whales could be taken.

In Europe, Basque fishermen were exploiting the whales of the Bay of Biscay by the sixteenth century and were also taking whales if they found them during fishing trips to the Grand Banks off Newfoundland. An estimate of the size of the trade is difficult because of a lack of records, but it is likely that the whaling fleet was never more than thirty or so ships, with an annual take of perhaps 300 whales (though with perhaps 100 more struck and killed, but not retrieved). The whale the Basques killed was probably the northern right whale, though they may also have occasionally found another species, later called the Greenland right whale. The whales were called 'right' because they were the right, i.e. the correct, ones to kill: they often swam close to shore and so could be towed relatively easily to shore stations for processing; they were slow and so easy to overhaul in a rowed boat (strangely, the slow swimming speed of the whale, a maximum of about

10km per hour (5 knots), contrasts sharply with their enthusiasm for breaching and other acrobatics); they were docile, and did not fight back when they were harpooned, though the flailing tail fluke of a wounded whale could break a man's bones or lift him out of a boat; they usually floated when killed (though it is estimated that 10–30 per cent of whales killed actually sank before they could be towed ashore: these whales were occasionally retrieved if they washed ashore or if decomposition gases later caused them to float); and they produced a huge quantity of oil from rendered blubber, this blubber mass being the reason they floated. Today the Greenland right is called the bowhead. It is a true Arctic species, but specimens are occasionally seen in more southerly waters.

The Basques hauled killed whales ashore to rendering stations on Newfoundland and Labrador, cutting the blubber into manageable lumps which were placed in three-legged boiling 'tri-pots' (or 'tri-works') from which the rendered oil was poured into barrels for transport back to Europe. There the oil – technically a liquid wax rather than an oil – was used to illuminate the homes of those that could afford it: whale oil was never cheap, and it seems that by the end of the sixteenth century northern right whales were becoming more elusive in the waters off Newfoundland. Whaling continued sporadically for another 150 years, mostly by American whalers operating from New England ports, but by the mid-eighteenth century the northern right whale was considered extinct in the Atlantic, as far as commercial hunting was concerned. At about this time whalers, chiefly American, but including ships from other nations, had begun to exploit the northern right whale population of the Pacific. Pacific whaling mirrored that in the Atlantic, particularly as better methods of meat preservation meant that meat as well as oil could be sold, making the whales an even more lucrative target. With the discovery of the Pacific population of bowhead whales, whaling in the Sea of Okhotsk and northern Pacific, where stocks of northern right whales were depleted, was followed by whaling in the Bering Sea, though later the whalers returned to southern waters as the stock of Bering Sea bowheads declined.

Today the population statistics of northern right whales make for depressing reading. It is estimated that there are perhaps only 300–400 whales in the Atlantic, with no more than three times that number in the north Pacific, chiefly in the Sea of Okhotsk. The whales are very vulnerable to being struck by ships, as they tend to ignore the sound of approaching vessels, and to becoming entangled in fishing gear. With such small populations, there are real fears that whale deaths due to these hazards may mean the species is doomed to extinction, particularly as females do not reach sexual maturity until they are 4–10 years old and give birth only every 2–4 years. Concerns over the fate of the northern right whale led US environmentalists to demand, in 2001, a speed reduction to 12 knots on shipping within 25 miles (40km) of ports on the eastern seaboard. The request was rejected by the US Government on the grounds of disruption of trade, a decision which led the environmentalists to sue the National Marine Fisheries Service (a sub-agency of the National Oceanic and Atmospheric Administration or NOAA) in 2005 for failing to protect a species which it (the Fisheries Service) recognised as critically endangered and the rarest of all the large whales. The northern right whale is listed as Endangered by the International Union for the Conservation of Nature and Natural Resources (IUCN) and under the US Endangered Species Act. In 2006 the NOAA reacted by proposing a 10-knot speed limit on specific routes for vessels longer than 20m (66 ft) during the calving season of the whales. In the event, the Final Environmental Impact Assessment, published in August 2008, suggested a series of six potential measures to protect the whales, including three possible speed limits (with a preferred option of 10 knots), the setting up of specific areas where the limits would apply on a seasonal basis and the possible rerouting of ocean traffic. The preferred strategy would come with a five-year 'sunset' period during which further research would be evaluated. As always, such extended periods for further research upset environmentalists who see an immediate problem, but the proposal is a major step towards species conservation. In addition to the NOAA proposal, there is also a proposal to set up an autobuoy listening programme in areas of known whale concentrations, the buoys detecting whales acoustically and forwarding position data to a central point from where it can be relayed to local shipping. It has also been discovered that although the northern right whale ignores vessel noise, it does react to high-pitched sounds, surfacing quickly. This might allow ships in sensitive areas to carry noise generators, though it would still, of course, be necessary for the ship to see a surfaced whale in time to avoid a collision.

In 1607 Henry Hudson set out from Britain in an effort to reach Cathay. Despite the fact that his name is now associated with the bay which takes a vast bite out of the northern mainland of Canada, and his fame is associated with the mutiny which saw him, his young son and other members of his crew cast adrift in an open boat, Hudson's 1607 voyage did not head north-west. At the time many scientists believed that the perpetual sunlight of the Arctic summer would melt the sea ice close to the North Pole. They reasoned that the ice which ships heading north encountered was actually a ring of ice which formed at the point where the extra sunlight of summer could not compensate for the biting winter cold. All that was needed was for a ship to look for a gap in the ice ring and to sail through: an easy trip across the sea around the Pole would then reach the ice ring again where a search would (hopefully) reveal another gap. Go through this and Cathay would be just a few days sailing away. Hudson could not find a gap in the ice ring (though the failure did not persuade the scientists of the shortcomings in their theory), but in the waters off Svalbard he found Greenland right whales, and he found them in huge numbers. Estimates vary, but it seems likely that as many as 25,000 whales swam the waters off Svalbard.

Despite the news of a new source of this most valuable of whales, it was two years before the British sent a ship to check on the validity of Hudson's account. The ship returned with confirmation and the British Muscovy Company immediately outfitted a ship and began commercial bowhead whaling in the waters off Svalbard and nearby Bear Island in 1610. To the astonishment of the whalers they found that in some bays of Spitsbergen the whales were so densely packed that they collided with ships at anchor or with anchor chains. Following the success of the 1610 expedition – which involved only a single ship – the Muscovy Company sent another, single, ship in 1611. In 1612 it sent two ships. In 1613 there were seven British vessels, and another seven from Holland, news of the discovery having spread. In 1614 the number doubled, and grew steadily in the years that followed.

So lucrative was the trade that problems arose between British and Dutch crews, with scuffles on shore and even battles on the sea, with cannon fire from both sides. So bad did the relationship become that the two governments intervened, drawing up an agreement which shared the whaling grounds between them. Ironically, in view of the British having discovered the whales, the Dutch had the better of the agreed terms. As with early Basque whaling, the whales were killed with harpoons from rowed boats, then towed to shore for rendering. The shore stations were set up annually, as it had proven impossible to avoid theft by early ships and the attempt to avoid the time-consuming job of rebuilding by men overwintering at the stations failed, as the men left behind succumbed to the rigours of the Arctic winter. The most famous station was the Dutch Smeerenberg (Blubber Town), whose remains can still be seen on Amsterdamøya, off Spitsbergen's north-western shore. Though it is still possible to find occasional references to Smeerenberg that include the idea of its having had a church, a bakery, a gambling hall, a dance hall and a brothel, serving a population of up to 10,000 (this idea was so firmly entrenched in the nineteenth century that even Fridtjof Nansen repeated it), the truth is more prosaic, archaeological evidence suggesting a population of 200 at most, housed in barrack-like rooms, and an absence of clergy, women and much else to raise the level of comfort above the very basic.

Danish, French and German whalers soon joined the British and Dutch in Svalbard waters and by the 1640s the population of bowheads had been so reduced that the boom was over. Whaling with fewer ships continued, but Smeerenberg was abandoned, probably in the 1650s, and within another decade or so, Svalbard had also been abandoned. But the bowheads could still be found in waters east of Svalbard and the Dutch set up shore stations on Jan Mayen. Eventually the whale stocks near Jan Mayen were also depleted, so that whaling was only possible by working the open sea, usually close to the ice edge, the rowed boats hauling dead whales back to a mother ship. The whale was then flensed beside the ship, and the blubber was packed in barrels for transport back to Britain or Holland. The hope was that the blubber had not become rancid before port was reached.

Soon the bowhead stock of the ice edge between Greenland and Svalbard was exhausted and interest turned to the Davis Strait and Hudson Bay, where the animals were still plentiful. But these waters were both a long way from Britain and Holland and were more dangerous, as ice conditions were more unpredictable. Another problem was the increase in American ships whaling in the area, which had the advantage of being much closer to home ports in New England and so were able to reach the whale grounds earlier. By the mid-eighteenth century the Americans had also perfected the use of on-board tri-works and so could render the whales immediately, reducing waste and volume. In Europe the need for lubricating oils and oil for lighting, and the demand for baleen – used for riding whips, carriage springs, fishing rods, parasol and umbrella ribs, shoe horns, and in corsetry – was increasing, and the sale of whale products was an important income for the emerging United States. With the depletion of the Davis Strait bowheads, the Americans concentrated on sperm whales, spermaceti, the oil from the whale's massive head, being much prized for candles, in perfumery and making up for the lack of oil a sperm whale yielded in comparison to a northern right whale. The production of oil and gas from coal in the UK, which caused a sharp fall in the price of whale oil, making the industry much less competitive, was a significant factor in the preservation of the sperm whale. But the introductions of these products came too late for the bowhead, whose eastern Arctic (i.e. Atlantic) population was severely reduced.

There were bowheads in the western Arctic too, and they had been hunted by early Eskimos for centuries, as the ritual site of Whale Alley and the remains in the Gambell boneyard testify. It is calculated that the western Arctic population of bowheads was about 30,000 by the 1840s, the native peoples on both sides of the Bering Sea taking perhaps 100 annually. Then, in 1848, Captain Thomas Roys took his ship, *Superior*, north in the Bering Sea in search of whales that he had heard about in Petropavlosk, the main port of Kamchatka. There stories were told of whales which were different from the (northern) right whales which whalers were used to hunting in the northern Pacific. Roys thought these new whales sounded the same as the Greenland right whales (bowheads) that had been hunted in the seas of the north Atlantic. He was correct. Close to the Diomede Islands, his ship became enveloped in fog, adding to the fears of his crew, already terrified to be so far north. When the fog cleared Roys saw whales in vast numbers and in all directions. From the first whale the ship's company took 120 barrels of rendered oil and the baleen was almost 12ft (3.8m) long. It was the most oil and the longest baleen the men had ever seen. At first they thought the whales were humpbacks, but it was now clear they were indeed bowheads. Roys took his ship through the Bering Straits to the Chukchi Sea. His men killed another eleven whales, from which they rendered 1,600 barrels of oil: the ship was full and Roys headed south back to his home port of Honolulu.

Other American whalers followed Roys' lead. In 1849 507 whales were taken in the Bering Sea with an estimated 70 more killed but not taken. In 1850 the number killed rose beyond 2,000. It was almost 2,700 in 1852, but thereafter declined steeply: by 1856 no bowheads were killed and the whalers had lost interest, sailing instead for the Sea of Okhotsk. They returned to the Bering Sea in the late 1850s when catches in the Sea of Okhotsk declined. But catches in the Bering Sea never again reached the heights of 1852 and the whalers were forced to explore further north, heading through the Bering Strait to the Chukchi Sea. In the 1870s the whalers explored eastwards into the Beaufort Sea. In 1889 they reached Herschel Island, where a wintering base was set up in 1890–1. Whaling in these northern waters was hazardous for the crews, with ships being lost or trapped in the ice: the whalers' graveyard on Herschel Island holds the remains of twenty-four men. Of these, five died in March 1897, when a sudden blizzard enveloped men playing baseball. In seconds a sunlit day with a temperature of 20°C had turned into a white-out at -20°C. The men could not find their way to cover, their frozen bodies being discovered the following day when the storm abated. But while this incident, and the loss of other lives and ships, was tragic, the whole story of whaling in the western Arctic is studded with tragedy. The whalers brought trade goods and alcohol, both of which disrupted the lives of the Inuit of Herschel Island and the northern mainland coast (particularly the Mackenzie delta), the trade goods leading to a loss of traditional skills and alcoholism leading to many deaths and murders. The whalers also brought measles, smallpox and syphilis: it is estimated that 90–95 per cent of the Herschel and Mackenzie Delta Inuit died of disease. Ultimately tales of debauchery of the Inuit forced the Royal Canadian Mounted Police to establish a base on the island in order to stop the worst excesses, though the base continued to operate until a combination of a flu epidemic and the lack of whales made it uneconomic to support. No bowheads were killed in 1912 or 1913. By then it is estimated that 18,600 animals had been killed (2,000 of those struck but not taken). A further 20,000 animals are likely to have been killed in the Sea of Okhotsk. Estimates for the kill in the eastern Arctic are more difficult as much of the killing was carried out before adequate records were kept. Some experts put the figure as high as 120,000. Today the bowhead is extinct (or very close to extinct) in waters east of Greenland. In the remaining waters of the eastern Arctic (the Davis Strait, Baffin Bay, Hudson Bay and Foxe Basin) the population is probably no more than a few hundred. In the Sea of Okhotsk the population is also in the hundreds. At such reduced levels the long-term survival of the species is questionable. In the Bering, Chukchi and Beaufort seas the number is estimated to be around 10,000 and to be increasing by about 3 per cent annually. This number was the subject of fierce debate in the aftermath of the Prudhoe Bay oil discovery, when native peoples acquired sufficient funds to contemplate whaling again, their take of bowheads having been effectively reduced to zero by commercial whaling activities. When they were denied an annual quota of whales because of the reduced stocks, the Yuppiat argued that there were more whales than researchers claimed. Using hydrophones to listen for whales the Yuppiat were able to establish that the figure was indeed higher than official estimates and on the Alaskan side of the Bering Sea were granted an annual quota. For the period 2008–12 this has been set at 280 whales for the Bering Sea as a whole (i.e both Alaskan and Russian native groups), with an annual maximum of 67, and a carry-over of 15 unused strikes.

Bowhead and northern right whale hunters also hunted grey whales, the bowhead hunters taking them during the summer when the bowheads had migrated beyond their reach. Grey whales had been taken by native peoples of both Asia (principally) and America for thousands of years, but were not taken commercially until the Japanese began catching them with nets in the seventeenth century. The annual take of the hunts, which ended in the nineteenth century, was probably no more than a few hundred. In America, commercial whaling began in Baja California in the 1840s, and about 11,000 animals were taken by the time the practice ended in the 1870s (though it continued from shore stations in California for another ten years or so). The Norwegians also hunted grey whales, taking them in the open sea in the 1920s to 1940s, though it is likely they took no more than 1,000 in total.

The bowhead whalers referred to the species as 'scrag whales' as they yielded only about 25 barrels of oil. They were also dangerous, acquiring the additional name 'devil fish' for their habit of attempting (and occasionally succeeding) to demolish the rowed boats of the whalers.

Despite the relatively small numbers, the grey whale kills were significant, as the population was never very numerous and the Californian whalers preferentially took pregnant females and females with calves as they were the easiest to pursue. Once the International Whaling Commmission (IWC) established a moritorium (see below), grey whale killing continued among the Yuppiat of Chukotka, with a kill of about 160 whales annually. Research showed that at this level of take the population was increasing. The population is now thought to be about 30,000, and at the same time that a native quota of bowheads was established, a quota of 620 grey whales was agreed for the same period, with a maximum of 140 in any one year. In 2008, 85 grey whales and 1 bowhead were taken by Chukchi natives. Quotas were also allocated for the same period to native Greenlanders: these included annual figures of 2 bowheads and 19 fin whales, as well as several hundred minke whales.

The bowhead is the largest of a trio of whales that are true Arctic dwellers. The narwhal is the most extraordinary of the three. These whales have only two teeth. In females these may not erupt during their lives, but in the male the left tooth (and very occasionally both) grow through the lip to form a tusk which was the likely source of the unicorn legend. The animals are confined to the eastern Arctic where they were hunted by native peoples (primarily the Inuit of eastern Canada), though never hunted commercially. After the introduction of rifles, snow scooters and outboard motors, overhunting by the Inuit reduced the population. The world population of the species is now thought to be about 35,000, though accurate information on numbers is difficult to obtain and possibly subject to large error. Trade in narwhal tusks is now governed by the Convention on International Trade in Endangered Species of Wild Fauna and Flora (CITES) and hunt quotas have been imposed by the Canadian Government. The Convention also seeks to protect other ivory-bearing animals, including the elephant and the walrus.

The third whale is the beluga, which has an essentially circumpolar range and was hunted both by native peoples and commercially. Commercial hunting was mainly confined to Svalbard. Russian whalers had hunted the beluga in the seventeenth century, but the main commercial trade was later, in the eighteenth and nineteenth centuries after the destruction of the bowhead population, when Norwegian whalers hunted the animals both in waters close to Svalbard and in the Barents Sea. The small size of the beluga, relative to the great whales, meant that the commercial trade was never enormous, and did not threaten species extinction. Today the circumpolar population of beluga is estimated to be about 200,000.

Although the northern right and bowhead whales were lucrative catches, the reason they were hunted was their slow swim speed, which allowed a rowed boat to overhaul them. American whalers hunted sperm whales from rowed boats as well, but the great whales, the rorquals, could not be hunted as they could reach speeds of 20 knots, far outstripping the boat of even a super-fit crew. But in the 1860s the Norwegian Sven Foyn invented the steam-powered whaler and the explosive harpoon, allowing the rorquals to be successfully taken. The invention ushered in whaling, on an industrial scale, of the whales of the Arctic fringe – blue, fin, sei, and humpback – though because of the population densities these were mainly taken in Antarctic waters. The slaughter of the great whales was such that it led to the creation of the International Whaling Commission (IWC) in 1946 with the intention of regulating whaling. In 1986, with some great whale species on the verge of extinction, the IWC called for a moratorium, though Japan objected and continued to hunt, largely for 'scientific' purposes. The science is considered dubious by many scientists, who point out that the number of whales killed is huge in comparison to the scientific papers published, and the papers themselves do not present data that could not have been obtained by non-lethal means. After the science is completed the carcass is sold to processing plants. The current position on whaling and the IWC is far from satisfactory. Iceland takes minke whales each year and in 2006 resumed killing fin whales. Norway also hunts on a small scale. Japan has also stated its intention to resume hunting humpbacks now numbers have increased (though it later bowed to US pressure and decided not to do so), and is widely accused of creating an IWC of its own choosing by offering inducements to non-whaling nations (some land-locked, some tiny Caribbean and Pacific islands) both to join the Commission and to vote with the Japanese. For Arctic species commercial whaling is unlikely in the future as stocks have not recovered sufficiently. However, the Arctic species are extremely vulnerable to pollution and increased shipping.

Walruses

In 1604, just a few years after Barents' voyage and the discovery of Bear Island, the Muscovy Company captain Stephen Bennet, in the *Speed*, reached Bear Island and found it occupied by a huge herd of walruses. His crew killed fifteen of them, and shipped walrus oil and ivory back to the UK. The following year Bennet returned and killed many more walruses. By 1606, when he came back again, his crew had become so good at killing the animals that the quantity of both oil and ivory Bennet took back had doubled. By 1613 walruses were extinct on Bear Island: the island has not been recolonized.

Though walruses had been killed by native peoples throughout the Arctic, the Bear Island expeditions were the start of the commercial exploitation of the animals. Though small in comparison to whales, walruses are large animals, and their tusks were extremely valuable. They also had the advantage of regularly hauling out on land so they could be hunted comparatively easily, and in reasonable safety. As long as whales were in ready supply, whaling always took precedence over walrus hunting, though there were always a few captains willing to settle for the less risky work of taking walruses as ivory fetched a high price in the world's markets. By the eighteenth century, whale populations were in decline and walruses became a more attractive prey, in both the western and eastern Arctic. Though they were much smaller than whales an adult walrus still provided a significant quantity of oil (about a half to two-thirds of a barrel per animal compared with 120–150 barrels from a bowhead. Butchering a walrus was also harder work than flensing a whale, but, of course, it was easier to kill a large number of walruses than to kill a whale). Walruses also provided ivory and hides, the latter being thick and both immensely strong and highly durable. In the nineteenth century, when sea mammal oil was no longer required for lubrication and illumination, and the market in walrus hide had also declined, walrus hunters frequently took only the tusks, abandoning the rest of the animal where it was killed.

The Atlantic walrus was never as numerous as, and has smaller tusks than, its Pacific cousin, probably amounting to no more than 10–15 per cent of the world population when the take was only by native peoples. Following the extinction of the Bear Island population the animal was hunted to near extinction on Svalbard (where the species, now fully protected, is on the increase) as well as in Canada's St Lawrence River. The Franz Josef and Novaya Zemlya populations were never commercially exploited (neither was the sub-species which is found in the Laptev Sea) and the largest eastern Arctic population, that of the Foxe Basin and Baffin Island, was not exploited until the 1920s, when as many as 175,000 animals were killed during the six-year period 1925–31. Hunting in Canada is now controlled under the Marine Mammal Regulations, with Inuit communities in the eastern Canadian Arctic being given annual kill quotas.

In the western Arctic the much larger population of walruses led to much higher killing rates, an estimated 3,500,000 animals being killed in the three centuries until 1960, when a Russian initiative led to the protection of the species across much of its range, Alaska setting up a sanctuary on the Walrus Islands in Bristol Bay.

The initial intense hunting period was by bowhead whalers in the 1870s. The ships took walruses (as with grey whales) at times when the bowheads had migrated out of easy reach (and were also in decline). Hunting was carried out on ice floes in the northern Bering Sea, and the technique was simple but dreadfully sad. Approaching a walrus herd on a floe, a boat full of hunters could approach so close that a hunter would ensure that a single first shot instantly killed an animal. A wounded animal would leap into the water, followed by the rest of the herd in panic, but if the first shot produced an instant death the other animals would look up, puzzled by the noise of the shot, then resume resting. The process could then be repeated, often until the entire herd had been slaughtered. (As a digression, the technique resembles the equally sad demise of many muskox herds to Norwegian fur trappers in north-east Greenland. Hunted by wolves, a muskoxen herd forms a circle, young animals at the centre, so that the wolves see only a line of massive heads and deadly horns. Only if the wolves can break the circle and make the muskoxen run can they make a kill. However, for a man with a rifle a herd of stationary muskoxen makes a thoroughly inviting target.) In the years between 1869 and 1878 records show that about 125,000 walruses were taken (with almost 36,000 killed in 1876 alone) and it is estimated that equally as many would have been shot and lost (a loss to take ratio of 1 to 1, though some estimates suggest a ratio as high as 3 to 1). That means that, of the total number of animals killed in 300 years, about 10 per cent were killed in just nine of those years. As a comparison, in the twenty-year period to 1820 1,600 *puds* of walrus ivory was exported from Russian Alaska to Moscow. The *pud* is an old Russian weight equivalent to 16.4kg (36lb), so the total was 26 tons of ivory. Taking an average of 1.5kg (3lb) for a pair of good tusks, this represents about 18,000 animals. (As a digression, the quantity of walrus ivory taken from the Bering Sea by the Russians was probably exceeded by the mammoth ivory extracted from northern Siberia: at its height, mammoth ivory extraction amounted to about 30 tons annually, with an estimated total extraction of perhaps 3,500 tons to date.)

As it is believed the population of Pacific walrus was probably never more than 200,000, the effect of the killing by the American whalers was dramatic, not only on the walrus population, but on the Yuppiat who depended on the animals for food. Starvation was widespread on the Diomede Islands, King Island and St Lawrence Island. In spring 1879 a whaling captain visiting St Lawrence island found that everyone in the three settlements he visited had died during the preceding winter. At another settlement there was just one survivor of a population of 200, while overall half the island's inhabitants had died. Despite the evidence of some of the whalers, and the testimony of the Yuppiat, American witnesses to the human tragedy were divided as to its cause. Many said that it was the introduction of alcohol to the native settlements, with the unwritten suggestion that drink and subsequent indolence was the chief reason. For many whaling captains the truth was glaringly obvious and they stopped the killing; others either disagreed or did not care and almost 7,000 walruses of the dwindling stock were killed in 1879, this number falling to 2,500–3,000 in the subsequent years (though apparently no walruses were killed in 1881). Thereafter, the number killed fell away to almost zero. Commercial hunting was finally prohibited in 1908. Walrus numbers then began to increase, but the population was again decimated by Russian hunters in the period 1935–55. It was this renewed period of slaughter which led to the Russian conservation initiative. In the United States the initiative spurred legislation and in 1972 the Marine Mammal Protection Act (MMPA) was passed. Although the Act did not apply only to walruses, it was this species that caused controversy as Alaska used it as the basis of a decision to ban hunting by native peoples in defined areas. This caused the native peoples to sue the US Department of the Interior for allowing the State such powers, on the grounds that the ban restricted their rights to use traditional hunting grounds. The conflict seems, though, to have been driven as much by other factors. The MMPA banned walrus hunting for trophies, but also restricted native hunting to that required for subsistence or for traditional crafts and demanded it should not be wasteful. In making these restrictions there seemed to be an underlying suggestion that the old tradition of using all parts of the walrus had gone, replaced by killing for the better parts of the animal, including the ivory. The suggestion was hardly likely to have been well received by the native peoples, particularly as it was becoming clear, not only here but across the Arctic, that younger generations were much less enamoured of the traditional ways of the native dwellers, a fact which was coupled with resentment of incomers over the fact that they had to buy meat at the supermarket while the native peoples were being granted exclusive rights to supplement their larders with 'free' food. In 1979 a US district court ruled that the native peoples need take no notice of State regulations, resulting in the State handing back responsibility for marine mammal management to the federal government. The Fish and Wildlife Service was tasked with management and immediately reimposed the State's hunting area ban. The Service, in conjunction with a native organization, now monitors the annual take of subsidence hunting. At present annual kills are 3,000–6,000 animals (for both Russian and American native peoples) from a population which is estimated to be again around 200,000.

In Greenland walruses occur on the north-west, west central and north-east coasts. Until the nineteenth century, they were taken chiefly by native Greenlanders, though European whalers operating in Greenlandic waters undoubtedly took some animals (though actual numbers are not possible to ascertain). During the nineteenth and early twentieth centuries, walrus kills in west Greenland remained steady at about 100–200 animals annually, but accelerated sharply in the 1930s as Greenlanders acquired boats and better weapons. Throughout the 1930s to 1950s it stood at 500–600 animals annually, before falling as the population declined. In north-east Greenland, where there was no indigenous population from the 1820s, foreign sealers began killing walruses in the 1950s and had soon depleted the population significantly. The result of these incursions, and the decline in population on the west coast, led to the introduction of regulations which meant that only Greenlanders were allowed to hunt walruses. In north-east Greenland, the lack of an indigenous population meant that the population has now recovered to pre-exploitation numbers (albeit a total of no more than about 1,000 animals). On the north-west and west coasts the rise in human population means that despite the regulations the walruses have not recovered and may well be in decline. There are estimated to be no more than 2,500 animals on the whole western side of the island, far fewer than in the nineteenth century. Worryingly, the annual take, of about 200 animals is considered by almost all authorities to be above the level of sustainability.

Birds

It is estimated that in 1845 there were over 100,000 significant bird colonies on Greenland. Today there are just a few thousand as a result of unmanaged shooting and egg collection. Protective legislation was put in place in 2001, but under pressure from Greenlanders for a restoration of their traditional right to hunt, these were cancelled in 2004. The pressure came not only from professional hunters who sell seabird meat as well as that from seals, but from ordinary Greenlanders, who wished to hunt, not for food, but for pleasure, or, as they put it, 'as a way of keeping in touch with their roots'.

Seals

All of the Arctic's seal populations have been exploited by native peoples, and in many areas they remain the most significant prey of these peoples. However, not all species have been hunted commercially (though in many cases, for instance ribbon and largha seals in the Bering Sea, this is as much to do with the limited population sizes and species range as any economic reason). Harbour and hooded seals have been hunted commercially, with significant depletion of their populations which has necessitated regulation of the hunts. By far the most visible of the hunts has been that of harp seal pups – 'whitecoats' – whose death by being beaten on the head with a club caused adverse publicity from the 1980s and has become a *cause célèbre* across the world on several occasions since. Photographs of the beautiful pups, their saucer-like black eyes sad and pleading against pure white fur, are set against shots of blood-stained ice and above tales of the animals being skinned while still alive. Despite annual public outcries in Canada, and the advice of scientists that the harp seal population is in decline, the Canadian government continues to subsidise the annual killing and has even raised recent quotas.

The killing of harp seal pups was also the cause of conflict between Norway and the emergent Soviet Union in the 1920s. The Pomors had taken pups in the White Sea for centuries, as had Norwegians (descendants of the Vikings rather than Sámi). (The Norwegians also periodically attacked Pomor villages, so the villages were placed above the first shallow rapids of local rivers, allowing the Pomors to escape into the local woodland in the time taken for the Norwegians to disembark.) By the 1920s the number of Norwegian sealing ships had increased to 200 and, after complaints from the Pomors, the Soviet authorities sent a diplomatic note to Norway requesting restraint. The Norwegians ignored the request and in the following year the Soviet authorities issued a decree allowing the Pomors to confiscate the ships and equipment of the 'poachers' and claiming the White Sea as internal waters. The Norwegians reacted by sending even more ships protected by the *Heimdal,* the former Norwegian Royal yacht, converted to an armed patrol ship. This decision was driven, in part, by the fact that Norway did not recognise the Soviet Union until 1924. The Soviets were outraged by this breach of their sovereignty, but lacking its own navy at that time was forced to compromise, and a concession of ninety Norwegian sealing ships was agreed. An uneasy peace lasted until 1928 when the Soviets arrested several Norwegian ships for poaching. The Norwegians appealed to the British to send naval vessels to protect its sealing fleet, and told the Soviets that the following year they anticipated that their sealing ships would be protected by two British cruisers. Stalin himself responded by sending a one word telegram: *zdem* – 'we are waiting'. There were no British cruisers, but the Soviets sent naval ships to Murmansk (creating the Soviet Northern Fleet). Another compromise was reached, but there were further problems in 1932 when the Norwegian frigate *Fridtjof Nansen* fired on Russian patrol boats. Finally, in 1936 the concession to Norwegian sealers was withdrawn. The harp seal population of the White Sea is thought to have been about 3,500,000 animals before the start of the 1920s overhunting. It is now thought to be about 2,500,000 and increasing. The Pomors still hunt the seals, but the numbers of Pomors involved is steadily decreasing. It is claimed that in 2008 about 10,000 seal pelts were burned as they could not be sold.

Hunting for Food

As well as hunting the species they sought to near extinction, the European fur trappers and whalers also hunted species for food, or traded goods with native peoples to obtain food from them. In places where the incomers concentrated this could have a devastating effect on local animal populations, which could lead to serious problems when the trapping or whaling ceased and the incomers moved on. The American whalers of Herschel Island took or traded caribou, muskoxen and seals for food. When whale numbers collapsed and the whalers left, the remaining Inuit of the island and mainland found that the dearth of animals meant starvation was added to the disease and alcoholism which the whalers had introduced. The depletion of another species must also be mentioned, though it was caused not by local hunting, but by slaughter along its migration route. The Eskimo curlew, a relatively small wading bird, was once abundant, breeding in the swamp lands of the Mackenzie delta. It migrated to South America in flocks of hundreds of thousands, flying through the Mississippi and Missouri valleys and across Texas, and all along the route it was relentlessly hunted. By 1916 the population had been so depleted that a hunting ban was introduced, but habitat destruction (the loss of prairies to agriculture along the migration route) meant no recovery occurred. The last credible sighting of the bird was in 1962 and it is now almost certainly extinct, a vanished Arctic species.

Northern native peoples also took fish to supplement their diet of terrestrial and marine mammals, and it is known that the Vikings fished the waters off Iceland and Greenland, but it was not until the medieval period, when the Basques and British fishermen began to exploit the cod of the Canadian Grand Banks, and later when the Norwegians and Russians fished the equally plentiful waters of the Barents Sea, that commercial pressures began to impact fish stocks. Cod appeared limitless, and the egg production of the female fish seemed high enough to ensure that no matter how many boats trawled waters of the northern Atlantic, cod would last forever. The same seemed true of the herring of the Barents and Norwegian seas, but man is an ingenious as well as a ruthless predator and soon found ways of denuding the stocks. By the second half of the twentieth century there were clear signs of overfishing.

In Iceland cod fishing was the basis of the economy and when, in the 1970s, the cod stock was clearly under threat Iceland responded by extending its coastal limit to 320km (200 miles) and fought 'Cod Wars', particularly with Britain, to confirm the limit, which initially was viewed as illegal by other nations. The limit has allowed cod fishing to continue in Icelandic waters (though the stocks of the fish elsewhere, such as in the North Sea, are now at dangerously low levels). Icelandic fishermen have, however, overfished local herring, and now pursue capelin, as well as pelagic fish. Icelandic and foreign fishermen have also fished the waters off eastern Greenland, catching Greenland halibut and cod, and some capelin. On the west Atlantic coast cod and halibut have also been fished, and shrimp fishing (a significant part of the Greenlandic economy) has been a major industry for many years. Stock reductions have again given cause for concern, but in general have not reached the worryingly low levels of those of the north-east Atlantic. Off eastern Canada overfishing has seen not only the virtual extinction of the cod, but also played a very large part in the actual extinction of the great auk.

In the north-eastern Atlantic overfishing has disrupted the normal migration patterns of herring, as well as causing a dramatic stock collapse. With herring catches falling, the fishing fleets of Norway and Russia transferred their attention to capelin and, predictably, stocks of that species also collapsed within a few years. After the collapse of the herring stock a total ban on fishing was enforced: remarkably, there had been no management efforts, or even any advice regarding the need for management before the stock collapsed, despite the warning signs from the overfishing of other species elsewhere and the disruption of the herring migration pattern. Stocks recovered sufficiently by the 1990s for fishing to begin again, this time with annual quotas defined by the Norwegians. In the Barents Sea, cooperation between Norway and the Soviet Union on understanding the fish resource began in the 1950s, with fishing regulated by a joint Commission set up in 1976. The Commission (now a joint Norway-Russia group) sets quotas for cod and haddock (which are shared 50/50 between the nations) and capelin (which is shared 60 per cent Norway, 40 per cent Russia). However, after cod stocks became critically low in the late 1990s, the Russian Commission members consistently opposed low quotas, apparently viewing quota specification not as an issue of stock management but as a conflict between the two countries, and even an East–West battle. This led to Norway reluctantly agreeing unsustainable catches, though by 2001 even the Russians had to admit that overfishing was endangering the species, and agreed to reduced quotas. However, in recent years Norway has become convinced that Russian trawlers are fishing illegally, with serious implications for the stock.

Fishing in the Bering Sea has come under intense scrutiny in recent years because fish stock reduction has been implicated in the decline of both sea otter and pinniped populations. Historically, fishing for salmon, halibut and Pacific herring has been important, with sport fishing for both salmon and halibut providing important tourist revenue, particularly in Alaska. Pelagic salmon fishing is also important to Russian and Japanese trawlers. Then, in the 1970s and 1980s the fishing of walleye pollock increased dramatically, with annual catches of over 2 million tons eventually being recorded. At the time, scientists considered the total weight of pollock in the sea as 10 million tons, and considered that

10 per cent was the maximum sustainable yield. After the years of high catches there was a sharp drop in yield, probably due to a combination of overfishing and a rise in water temperature, though experts are divided on which of these was the more significant. The populations of northern fur seals, sea otters and Steller's sea lion declined dramatically at the same time, the sea lion by 80 per cent overall and by 90 per cent in Russian waters. (Ironically, in waters off British Columbia the Canadian authorities agreed a cull of the sea lions as they were accused of endangering fish stocks.) One likely scenario sees warming water pushing temperature-sensitive organisms at the bottom of the food chain north, and crustaceans dying as they are unable to respond fast enough. Young fish also die from lack of food, and with fishing reducing adult fish numbers, populations quickly decline. Next the sea lion population slumps because of a lack of fish. Then, perhaps, orcas, deprived of the mainstays of their diet – fish and sea lions – take otters instead. The otters are the main predators of the sea urchins that graze on kelp. The subsequent increase in urchin numbers then causes the kelp forests around the Aleutians to thin or disappear. Many authorities believe that the ecology of the southern Bering Sea has been irreparably altered; as this sea is one of the wonders of the sub-Arctic, the loss of this ecosystem, which supports, among other things, millions of seabirds, would be a tragedy.

The Russians and Americans have come into conflict over fishing in the sea, particularly over the position of the boundary line between the two countries (see Chapter 4 for the history of this conflict). Russian trawlers regularly fish at the boundary line, taking pollock as the fish migrate from American towards Russian waters, US stock managers being convinced that Russia is consistently taking tonnages that threaten the stock. Attempts by the United States to police disputed waters have been met by very aggressive responses from Russian trawlers, with threats to ram US Coastguard vessels. At present no resolution of the conflict appears imminent, and overfishing of the Bering Sea remains a problem. The American pollock quota for 2009 is 815,000 tons (which would be the lowest yield since records began in 1977). This figure pleases no one, being considered too low by fishermen and too high by conservationists. The figure was derived to allow a partial recovery so that the quota could rise to 1,200,000 tons in 2010, a rise which again pleased no one.

Two other effects of fishing cannot be ignored. Bycatch, the name given to the accidental taking of sea mammals and birds caught in, and killed by, fishing tackle, is a significant cause of mortality of some species. It is, for example, considered a real threat to albatross species in the North Pacific, particularly the endangered short-tailed albatross. The second effect is damage to the seabed by trawl nets. In shallow waters with soft seabeds it is likely that storms cause more disruption to sediments than trawling, and that benthic creatures have adapted to occasional disturbance. In deeper waters where the effects of storms are negligible, disturbance from the dragging trawl nets will have more impact. This impact increases on harder seabeds, with considerable damage being caused to corals and rocky environments. Steller's sea lions are also victims of drift nets: at rookeries in the Sea of Okhotsk up to 10 per cent of adult animals have 'collars' around their necks made of drift netting. But these 'collared' individuals represent the tip of the iceberg of the mortality, as it is impossible to estimate number of animals killed in the bycatch.

The *Manhattan* on her controversial voyage through the North-West Passage in 1969.

Chapter Four
International Law in the Arctic

> There is a growing national and international appreciation of the importance of Arctic ecosystems and an increasing knowledge of global pollution and resulting environmental threats. The Arctic is highly sensitive to pollution and much of its human population and culture is directly dependent on the health of the region's ecosystems. Limited sunlight, ice cover that inhibits energy penetration, low mean and extreme temperatures, low species diversity and biological productivity and long-lived organisms with high lipid levels all contribute to the sensitivity of the Arctic ecosystem and cause it to be easily damaged. This vulnerability of the Arctic to pollution requires that action be taken now, or degradation may become irreversible ... Therefore, this Strategy should allow for sustainable economic development in the north so that such development does not have unacceptable ecological or cultural impacts ... The implementation of the Strategy will be carried out through national legislation and in accordance with international law, including customary international law as reflected in the 1982 United Nations Convention on the Law of the Sea.
>
> **Arctic Environmental Protection Strategy, signed June 1991 by Canada, Denmark, Finland, Iceland, Norway, Sweden, the USA and the USSR**

In 1959 the Antarctic Treaty (Appendix 2) was signed by seven nations with territorial claims on the continent – Argentina, Australia, Chile, France, New Zealand, Norway, and the United Kingdom – together with Belgium, Japan, South Africa, the Union of Soviet Republics (USSR) and the United States of America (USA). Since 1959 many other nations have also signed the Treaty. The Treaty had multiple aims, including the limiting of activities on the continent to scientific investigations, and the provision that all activities should be peaceable. Perhaps most importantly, signatories of the Treaty agreed that no new territorial claims would be made on the continent and that existing claims would not be increased. This provision was necessary as the claims of Argentina, Chile and the UK overlap (the UK claim entirely overlapping the Argentine claim, and the Antarctic Peninsula being claimed by all three nations). About 16 per cent of the continent is also unclaimed by any nation. Perhaps surprisingly the Treaty makes no mention of environmental protection, though a series of conventions signed since 1959 have rectified this position (see also Appendix 2). In particular, the Protocol of 1998 prohibits the exploitation of Antarctic mineral resources (for other than scientific purposes). Sovereignty claims on the continent are both difficult to establish and maintain, but as long as exploitation is prohibited this does not represent a serious issue, even for the conflicting claims of Argentina, Chile and the UK.

The success of the Antarctic Treaty in providing a single, encompassing agreement signed by both territorial claimants and other nations leads the casual observer to the view that such an agreement should be equally possible in the Arctic. But there are profound differences between the two polar areas. Antarctica is a continent surrounded by a violent ocean. It is uninhabited and has never been inhabited. The Arctic is an ocean, part permanently frozen, part seasonally frozen, almost entirely encircled by land masses which are largely inhabited and which have been inhabited since prehistory. For each of the eight Arctic nations – Canada, Denmark (Greenland), Finland, Iceland, Norway, Russia, Sweden and the United States (Alaska) – the Arctic section is the far north of their country, while their major interests (until very recent times at least) concern the population and resources of the southern section.

Added to this sense that the Arctic region was of lesser importance, especially since development of a cold, hostile land and an ice-clad sea seemed both difficult and not worthwhile, was the Cold War, which made peaceful agreements on northern issues of common interest problematic. Because of the proximity of the USSR and USA across the Bering Sea, together with the shrinkage of distances between all regions of the two countries at high latitudes, the Arctic was always going to become a strategically important area in the wake of the 1939–45 war. The development of nuclear-powered submarines which were able to travel for long distances and durations beneath the sea ice of the Arctic Ocean, using the ice as a shield against spotting from the air, while at the same time being able to travel within easy striking distance of potential targets, meant the Arctic became even more strategically important in the post-war era. Strategic importance did not necessarily mean that agreements were impossible, but with the USSR and USA – and NATO and the Warsaw Pact nations – being locked in positions of wariness or muted hostility, bilateral and multilateral agreements became difficult.

They were not, however, entirely absent and some actually predated the Antarctic Treaty. In 1911 the Convention on the Preservation and Protection of Fur Seals was signed between Japan, Russia, the UK (on behalf of itself and Canada) and the United States. The convention was necessary to avoid the extinction of the northern fur seals which was being overhunted at its breeding grounds in the Bering Sea. As less hostile relations developed between East and

West, further agreements were signed, notably those between adjacent Arctic nations on the boundary lines between them. Of perhaps greater significance was the 1973 agreement on the protection of polar bears and their habitat (Appendix 3) between Canada, Denmark (Greenland), Norway, the USA and the USSR. As noted in Chapter 6, the Agreement has a limitation which has become increasingly important with the passage of time, but can also be seen as the beginning of cooperation between the Arctic nations.

Some, perhaps, would argue that the impetus for Arctic agreements came not only from a thawing of the Cold War and a recognition of common interests (such as the welfare of the polar bear), but also from the realization that the Arctic could be reached in relative comfort, giving hope that development could be possible at some future time. The rescue by air of the passengers and crew of the *Chelyuskin* from the East Siberian Sea in 1934 had been the first indication. Trapped by sea-ice during an attempted transit of the North-East Passage the ship was crushed and, eventually, sank. But the sinking was a protracted process and all on board were able to get off and to set up a camp on the ice from where they were rescued by a series of flights which landed on a makeshift, icy runway. The pilots involved were the first recipients of the award of Hero of the Soviet Union. The relative comfort enjoyed by the marooned travellers, and the relative ease with which an acceptable runway could be engineered, encouraged the Soviets to set up drift stations in the High Arctic, manned camps on the ice, which drifted with the ice currents of the Arctic Ocean while scientific observations were carried out. The first such station, manned by four men (two of whom had been on the *Chelyuskin*) and a dog named Veselyi (Happy) was set up in June 1937, an aircraft landing the party 25km (15½ miles) from the North Pole. The camp drifted until February 1938 when the party was rescued from a melting floe close to Scoresbysund in east Greenland.

In April 1948 a Soviet aircraft landed at the North Pole – the occupants becoming the first people to have unquestionably stood there – while ten years later the US nuclear submarine *Skate*, surfaced at the Pole. The USSR nuclear-powered ice-breaker *Arktika* reached the North Pole in August 1977, the first surface vessel to do so. There are now regular trips in the summer using the Russian nuclear-powered *Yamal*, as well as the newer ships *Sovetskiy Soyus* and *50 Let Pobedy*, which are regularly chartered by western travel companies.

Irrespective of which reason is considered to be paramount, ease of travel, the possibility of development and a growing awareness of the fragility of the Arctic and its wildlife all contributed to a growing number of agreements on Arctic issues in the decade from 1973. There are now several dozen such agreements and from some contentious issues have arisen. One specific example is the creation of the International Whaling Commission from the 1946 International Convention for the Regulation of Whaling. While not specific to the Arctic, it has critical implications for the area as it is rich in whale species and numbers, and has a history of whaling, both by native peoples and by Arctic nations (Iceland and Norway in modern times).

Mikhail Gorbachev's 1987 Murmansk Speech

The next major initiative on Arctic cooperation came in October 1987 when the then Secretary-General of the Soviet Communist Party, Mikhail Gorbachev, made a speech in Murmansk which is now widely seen as a crucial one, not only in the development of the Arctic, but in ending of the Cold War. The speech was also critical in the establishment of *glasnost* (openness, transparency) and *perestroika* (restructuring) in the USSR, which ushered in reforms which saw the fall of the Berlin Wall and the break up of the Soviet Union. The Murmansk speech can therefore justifiably be seen as one of the most important of the twentieth century.

With regard to the Arctic Gorbachev spoke of his desire to see a 'Zone of Peace' established. In particular he proposed that a nuclear weapons-free zone should be declared in Northern Europe; that restrictions should be placed on naval activities in Arctic seas; that peaceful cooperation should be the basis for the development of the area's resources; and that there should be cooperation between Arctic nations on issues such as science, environmental protection and the rights of native peoples. Gorbachev favoured northern science as it had, he said, 'great significance for all mankind'. Given that the Arctic is now seen as the litmus paper for the Earth as far as climate change is concerned (see Chapter 6), this was a remarkably prescient suggestion.

Gorbachev's Arctic theme was taken up later by other Soviet leaders. One example was Alexei Rodionov, the Soviet Ambassador to Canada, who, in 1989, declared that the Arctic nations should be 'expanding cooperation in the field of Arctic natural resource development; conducting scientific research, including research aimed at keeping an ecological balance in the region; and ensuring the social and economic rights of native populations.' It was a wonderful idea, and had it been taken up more enthusiastically it might have had a bearing on the Arctic's future. But the fine words were before the phrase 'climate change' had become common currency, and before Vladimir Putin's post-Soviet Russia and George W. Bush's United States had decided that the race to develop the Arctic outweighed consideration of its future or, indeed, the future of the Earth.

The Arctic Environmental Protection Strategy

In the same year that Rodionov was advancing Gorbachev's Arctic ideas, a letter from the Finnish government was arriving in the mail boxes of the seven other Arctic nations. The initiative was prompted by a realization that 'time was running out in the protection of the Arctic environment'. While the ecology of all of Earth's biomes is vulnerable in some ways, that of the Arctic is particularly fragile. The low temperatures the area experiences slow the breakdown of substances, both those that naturally occur and the pollutants brought in directly by man, or carried by wind and water; the harsh climate also slows the regeneration time of plants, the growing season being short; the populations of animal and bird species, are often more concentrated than in temperate zones, despite many species having circumpolar ranges, which makes them more vulnerable to catastrophic events; and the ice-bound nature of the area makes clean-up after catastrophes more difficult, and so protracted, and expensive. In addition, the growing awareness of the effect of increasing carbon dioxide levels in the Earth's atmosphere, was leading to an understanding that positive feedback mechanisms were causing the Arctic to see higher, and perhaps faster, temperature rises than elsewhere. All the issues in this list are trans-border, i.e. they are very likely to effect all Arctic nations irrespective of, for instance, where pollutants arise.

Finland suggested a meeting of the eight Arctic nations, and in September 1989 they gathered in Rovaniemi. Successive meetings in Yellowknife, Canada and Kiruna, Sweden, led to a final meeting back in Rovaniemi in June 1991 at which the Arctic Environmental Protection Strategy (AEPS) was adopted. The Strategy committed the signatories to:

a) Cooperation in scientific research to specify sources, pathways, sinks and effects of pollution, in particular, oil, acidification, persistent organic contaminants, radioactivity, noise and heavy metals as well as sharing of these data;
b) Assessment of potential environmental impacts of development activities;
c) Full implementation and consideration of further measures to control pollutants and reduce their adverse effects to the Arctic environment.

The signatories also committed themselves to assessing 'on a continuing basis the threats to the Arctic environment through the preparation and updating of reports on the state of the Arctic environment, in order to propose further cooperative action.' They further committed themselves to implementing the following measures of the Strategy:

a) An Arctic Monitoring and Assessment Programme (AMAP) to monitor the levels of, and assess the effects of, anthropogenic pollutants in all components of the Arctic environment. To this end, an Arctic Monitoring and Assessment Task Force was to be established.
b) A Protection of the Marine Environment in the Arctic (PMEA) initiative, to take preventive and other measures directly or through competent international organizations regarding marine pollution in the Arctic irrespective of origin.
c) An Emergency, Prevention, Preparedness and Response (EPPR) initiative, to review the adequacy of the geographical coverage of the Arctic regions by cooperative agreements with particular reference to actions in response to significant accidental pollution from any source, coordination and harmonization of preventive policies, and establishment of a system of early notification in the event of significant accidental pollution or the threat of such pollution.
d) A Conservation of Arctic Flora and Fauna (CAFF) working group to exchange information and coordinate research on species and habitats of flora and fauna in the Arctic. CAFF considers the practices of Arctic states with respect to conservation and the management of Arctic species, and the relationship to and use of such species by native groups.

Despite knowledge of the greater risk climate change posed to the Arctic, it is a notable omission from the list of possible hazards the Arctic environment was thought to be facing. However, as it has evolved, the work of AMAP (which has a Norwegian secretariat) has included consideration of the effects of climate change on Arctic ecology.

The AEPS is in many ways an extraordinary document, and can be viewed as an equivalent to the Antarctic Treaty, but for the international lawyer it has significant problems. In particular it does not define the Arctic, allowing each signatory State to use its own definition, and its use of 'plain English', though welcome to the general reader, is at odds with the requirements of a legal treaty which defines rights and obligations on the signatories. It is also the case that AEPS did not require ratification by the countries party to it: usually ratification is required to produce a binding agreement.

Before the Finnish initiative, in both the 1970s and 1980s, the Canadians had several times proposed the setting up of an Arctic Council. With AEPS in place, the idea of a Council developed naturally, and finally came into being in Ottawa in 1996. It was established as a high level intergovernmental forum as a means of promoting cooperation and coordination between the eight Arctic countries, with the involvement of native groups and Arctic dwellers of non-native origin. The aim was to consider all issues common to the Arctic nations, but in particular sustainable development and environmental protection. The Arctic Council now includes the eight Arctic states, as well as permanent participants from the Aleut International Association, the Arctic Athabaskan Council, the Gwich'in Council International, the Inuit Circumpolar Council, the Sámi Council and the Russian Arctic Indigenous Peoples of the North. The Chairmanship of the Council rotates every two years, and was held by Norway, which took over from Russian Federation, until March 2009. Having common aims, the three Nordic countries (Norway, Denmark and Sweden) have agreed to a joint chairmanship until 2012, though technically Denmark and Sweden will each hold the chair for two years. The Arctic Council is administered by a secretariat based in Tromsø, Norway. The Arctic Council has now taken over the administration and coordination of the four working groups set up by AEPS, and has added two more, the Arctic Contaminants Action Program (ACAP) and Sustainable Development Working Group (SDWG). ACAP was set up to develop and implement an integrated hazardous waste management system in the Arctic, while SDWG was created to protect and enhance the economies, culture and health of the inhabitants of the Arctic, in an environmentally sustainable manner. But, as with the AEPS itself, the Arctic Council has no legal status. This is considered advantageous by the member countries as it can aid governments to comply with domestic legislation and manage foreign policy without hindrance. It also facilitates funding, as there is nothing to prevent any country contributing funds as a political, rather than legal, gesture, thus allowing funding to be terminated easily and without legal consequence. The programs of the Arctic Council are funded voluntarily by individual Arctic states. Under current practice, states propose projects or identify working groups they wish to support, and those governments that are interested take the lead in implementing and paying for them. Thus, for example, Norway pays for a secretariat for AMAP, Iceland for PMEA, the United States and Iceland for CAFF, and Denmark provides most of the funding for an Indigenous Peoples' Secretariat located in Copenhagen. The lack of legal status also allows contentious issues to be raised and discussed without the need for public announcements that are either written by consensus and, as such, so bland as to be essentially worthless, or written by the majority and therefore cause embarrassment (and possibly withdrawal) of a member country.

Yet while it is easy to see why these reasons have been arrived at, and even to sympathize with them, the fact is that the lack of legal status and the lack of a need to commit to potentially unpopular decisions mean that the Council can be seen as toothless, a talking shop with no powers to prevent any individual country from pressing ahead with development in the Arctic if its perceived national interest outweighs the strictures of the AEPS. Recent events suggest that this analysis is not so far wrong.

The Law of the Sea

While agreements between nations are clearly useful, they do not have the status of law and so do not impose specific legal obligations on the countries concerned. For the layman the difference can be not only difficult to grasp, but may seem esoteric, particularly when the language used to describe international agreements can be identical, yet mean different things in different contexts. Examples are the Montreal Protocol on Substances which Deplete the Ozone Layer, which is a treaty and therefore covered by international law, while the Kyoto Protocol is not actually a treaty, rather an aspiration, a rule of diplomatic etiquette as it were.

However, the absence of globally recognized international agreements does not mean that neighbouring Arctic states are necessarily left in the murky waters of legislative chaos. All Arctic countries have bilateral agreements of some sort, though these may not be recognized by other countries, including neighbouring states. A very good example of an international agreement which does not have the status of international law is the 1988 Agreement between the United States and Canada on Arctic Cooperation. The aims of the Agreement were the encouragement of shared interests and security, one specific point being on the use of ice-breakers in Arctic waters. The Agreement noted the advantages of developing cooperative measures for ice-breaker navigation in the respective waters of the two nations, and the need to share information gathered from and relating to ice-breaker voyages (something which the Agreement says will be done 'in accordance with international law'). The Agreement then notes that the voyages of US ice-breakers 'within waters claimed by Canada to be internal (will) be undertaken with the consent of Canada.' (Agreement Clause 3). Agreement Clause 4 states that nothing in the Agreement 'affects the respective positions of the Governments of the United States and of Canada on the Law of the Sea.' In other words, the Agreement does require a legal obligation from either country, each remaining free to pursue an interest which legal status would deny it. This particular Agreement, and the freedom it gives each country, remains crucial in discussions on the status of the North-West Passage, as we shall see below.

The 1988 Agreement is also interesting for covering Arctic waters. As the Arctic is an ocean – though one which is both permanently and seasonally ice-covered, at least for the present – it is, in principle, covered by the Law of the Sea. The proviso is sea ice, international law struggling with the status of ice in general and an ice-covered sea in particular: sea ice is subject to drifting, and the drift camps the Soviets set up on the Arctic Ocean sometimes covered great distances over long time periods. For the lawyer schooled in land or ocean masses, things that move about in space and time are a nightmare.

Ice on land (glaciers and ice caps) is fine – it is the equivalent of land. Ice on the sea is a problem. Though the scientist defines numerous forms of ice found on the sea, there are essentially only three forms – shelf ice, fast ice and pack ice. Shelf ice occurs when a glacier reaches the sea and, rather than calving icebergs at the interface of land and sea, is of sufficient thickness that it continues to flow as a homogenous mass across the water. Shelf ice is almost exclusively found in Antarctica where the glaciers flowing down from the continent's high plateau have sufficient volume for it to form. Historically there have been ice shelves in the Arctic – on northern Ellesmere Island, Franz Josef Land and Svalbard's Nordaustlandet – but these were always small (by comparison to Antarctic shelves) and have all but disappeared. Fast ice is frozen sea which is attached to (i.e. 'fast to') land. Pack ice can be used as a term encompassing the various forms of sea ice encountered away from land. The breaking of the seaward edges of shelf ice gives rise to the vast tabular icebergs of the Southern Ocean. Such icebergs have occurred from the limited ice shelves of northern Ellesmere Island. Called ice islands, they do not escape the pack ice of the Arctic Ocean and have occasionally been used as platforms for scientific research as they are stable. Icebergs in the northern hemisphere, such as the one which ended the brief career of the *Titanic*, arise from the calving of 'tidewater' glaciers (i.e. glaciers that reach the sea) rather than the shelf ice break-up. Such icebergs are smaller and an irregular shape making them very unstable as erosion by wave, wind and sun often causes them to tip.

From the point of view of a lawyer, all three forms represent a problem as they can be permanent, or reasonably so. Shelf ice is usually assumed to be land (and is treated as such in the Antarctic Treaty), and in a case against an Inuit who had committed a crime while on fast ice the Canadian courts declared that they had jurisdiction despite the transient nature of the ice as, at the time, it had the 'attribute of land' and as such was part of Canada. The Inuit was found guilty and sentenced accordingly. Sea ice represents a much more difficult problem (though it is worth noting that the United States successfully prosecuted Mario Escamilla for a killing that took place in 1970 on an iceberg called Fletcher's Ice Island (also known as T3) which at the time was at 84°47′N, 106°28′W, north-west of Ellesmere Island).

Equally problematic is the fact that 'high seas' as a definition implies both the right to, and the ability to, navigate. On the ice-covered ocean navigation is all but impossible, except at the ice fringes, to anything other than an ice-breaker and a very powerful ice-breaker at that. How does the law of the sea, constructed to deal with navigation by surface vessels apply to an area where surface vessels cannot go, but submarines have free access?

With climate change causing a rapid decline in both seasonal and permanent ice coverage that difficulty is lessening. But the access the lack of ice allows both hinders and helps the lawyer. It helps with an understanding of the Arctic Ocean, but makes life more difficult when defining the status of those large tracts of the Arctic which have no settled population and, indeed, are very hostile as regards permanent settlement. This means that laws written for application to the 'normal' position of well-defined national boundaries may not apply.

Open water, as now covered by international law, began as an assumption of freedom of navigation on the sea, a freedom that has been eroded over time by the extension of the influence coastal country's decided to allot themselves over adjacent waters. The maritime limit has been extended from essentially no distance in medieval times to a standard of three nautical miles (1 nautical mile = 1.852km, and was originally defined as one minute of latitude along any meridian), which has been increased to 200 nautical miles (370.4km) as concerns over fishing resources grew. The initiative for the change was unilaterally imposed by Iceland after the Cod Wars of 1950 and 1970, but it is worth noting that in 1945 the US President Truman extended American territorial waters to cover the 'continuous shelf zone', which may be seen as starting the extension 'race'. However, the United Nations Convention on the Law of the Sea (UNCLOS), agreed in 1982 (but not coming into force until 1994) standardized the limit at 12 nautical miles (22.2km) for the purpose of defining the border of the 'high sea'. Within the 12 nautical miles are 'territorial waters'. Territorial rights include the airspace above those waters and the seabed below them.

Within territorial waters the coastal nation is free to set laws, exploit resources and regulate usage by foreign vessels. However, such vessels are accorded the right of 'innocent passage', this being defined as continuous and speedy transit. Not surprisingly, spying and weapons practice by naval vessels do not constitute innocent passage – but neither does fishing. Submarines are required to travel on the surface and show their national flag. The right of innocent passage in territorial waters can be suspended if a country claims that such suspension is essential to its security.

In general the 12 nautical mile limit has been observed, though a particular problem exists for channels between an island and the mainland of another country which are less than 12 nautical miles apart.

UNCLOS also defines Exclusive Economic Zones (EEZs), which extend 200 nautical miles from the defined baseline. Within the Zone a country has exclusive rights over the exploitation of natural resources, including fishing and, as an example, the erection of oil-drilling platforms. Other countries have the right to lay submarine pipes and cables in EEZs, and the right to navigate with them, and to overfly them. Coupled with the EEZ is the definition of the continental shelf, the natural, sub-sea extension of the continental margin. A country's continental shelf is defined as a minimum of 200 nautical miles (i.e. is equivalent to the EEZ for those margins which do not extend as far as 200 nautical miles) but may extend further. However, there are defined limits for extensive shelves: they must not exceed 350 nautical miles, or 100 nautical miles beyond the 2,500m isobath (the line linking depths of 2,500m). Coastal countries have exclusive rights to exploit the resources on or beneath their shelf, but not to creatures which live in the waters beyond 200 nautical miles.

Despite these apparently very tight definitions, there are problems. Arctic nations have claimed certain off-shore waters as 'historic waters', even as 'internal waters'. Internal waters are those which lie on the landward side of a low-water baseline. That would seem to exclude anything which a casual observer might classify as sea, but for deeply indented coasts, islands very close to shore or where the coastline is unstable (for instance at deltas), a straight baseline is allowed, rather than the low-water line, so that certain (usually small) areas of sea become 'internal waters'. Foreign vessels have no right of passage through such waters. The most famous example of the application of straight baselines was in a dispute between Britain and Norway over fishing close to Norway's northern coast: in 1951 the International Court of Justice upheld a 1933 decision that Norway was entitled to draw straight baselines to define the limit of its 'coast' because of the fjords which bit deeply into its mainland.

The idea of historic waters represents a legal grey area. They are defined as waters over which a country has a legitimate claim to ownership based on prolonged and essentially sole usage, coupled with the acceptance of the position by other countries. While the principle exists, and has even been recognized in law (including UNCLOS), actually proving the existence of such waters is difficult. There have been cases in which bays have been recognized as such, but they are a special case as they are invariably enclosed, to an extent, by the claiming country. For waters that are distant from the shore, the situation is much more difficult. For Canada to claim the waters of its Arctic archipelago, or for the Russians to claim the waters between its mainland and the near-shore archipelagos as historic, it would be necessary for them to show that there has never realistically been a right of passage through them. Unfortunately in each case that position is difficult to defend, particularly as the understanding of the channels and seas of Arctic

Canada owes much to the voyages of foreign expeditions, while the North-East Passage was first traversed by a Swede, the third transit being by the Norwegian Roald Amundsen.

A further problem exists in defining the Arctic Ocean itself. Article 122 of UNCLOS defines an enclosed or semi-enclosed sea as 'a gulf, basin or sea surrounded by two or more States and connected to another sea or the ocean by a narrow outlet or consisting entirely or primarily of the territorial seas and exclusive economic zones of two or more coastal States.' But does the Arctic Ocean constitute a semi-enclosed sea given that it is not truly connected to other seas only by a narrow outlet, nor is it either entirely or primarily either a territorial sea or encompassed by EEZs? This seems another esoteric argument, but has enormous relevance as Article 123 of UNCLOS, on the 'Cooperation of States bordering enclosed or semi-enclosed seas' says that such States

> should cooperate with each other in the exercise of their rights and in the performance of their duties under this Convention. To this end they shall endeavour, directly or through an appropriate regional organization:
> (a) to coordinate the management, conservation, exploration and exploitation of the living resources of the sea;
> (b) to coordinate the implementation of their rights and duties with respect to the protection and preservation of the marine environment;
> (c) to coordinate their scientific research policies and undertake where appropriate joint programmes of scientific research in the area;
> (d) to invite, as appropriate, other interested States or international organizations to cooperate with them in furtherance of the provisions of this article.

Even more important is the legal position of the deep seabed. This is also covered by UNCLOS which defines 'The Area' as the seabed, ocean floor and subsoil thereof which lies outside the EEZ of the countries which border the sea. In the case of the Arctic that limits 'The Area' to the bed of the Arctic Ocean lying outside the EEZs of the United States, Canada, Denmark/Greenland, Iceland, Norway and Russia, though the situation is complicated by the claims of both the UK (representing Canada (1925)) and Russia (1926) to sectors which extend all the way to the Pole (see Chapter 2 on the 'sector' principle).

Part XI of UNCLOS states that 'The Area and its resources are the common heritage of mankind.' Defining resources as 'all solid, liquid or gaseous mineral resources *in situ* in the Area at or beneath the seabed, including polymetallic nodules'. Part XI goes on to state (Article 137):

1. No State shall claim or exercise sovereignty or sovereign rights over any part of the Area or its resources, nor shall any State or natural or juridical person appropriate any part thereof. No such claim or exercise of sovereignty or sovereign rights nor such appropriation shall be recognised.
2. All rights in the resources of the Area are vested in mankind as a whole, on whose behalf the Authority shall act. These resources are not subject to alienation. The minerals recovered from the Area, however, may only be alienated in accordance with this Part and the rules, regulations and procedures of the Authority.
3. No State or natural or juridical person shall claim, acquire or exercise rights with respect to the minerals recovered from the Area except in accordance with this Part. Otherwise, no such claim, acquisition or exercise of such rights shall be recognised.

In Article 140 of Part XI, the idea of shared resources is expanded:

1. Activities in the Area shall, as specifically provided for in this Part, be carried out for the benefit of mankind as a whole, irrespective of the geographical location of States, whether coastal or land-locked, and taking into particular consideration the interests and needs of developing States and of peoples who have not attained full independence or other self-governing status recognised by the United Nations in accordance with General Assembly Resolution 1514 (XV) and other relevant General Assembly resolutions.
2. The Authority shall provide for the equitable sharing of financial and other economic benefits derived from activities in the Area through any appropriate mechanism, on a non-discriminatory basis, in accordance with Article 160, paragraph 2(f)(i).

It includes a general statement on development (Article 141) that 'The Area shall be open to use exclusively for peaceful purposes by all States, whether coastal or land-locked, without discrimination and without prejudice to the other provisions of this Part.'

Implementation of Part XI of UNCLOS is overseen by the International Seabed Authority (ISA) which was established in 1982.

The US objected very strongly to Part XI, chiefly because, it claimed, the provisions represented a threat to its economic interests and national security. It also considered the provisions were anti-capitalist, acting against the free market to the advantage of communist states (meaning China at that time). It therefore declined to ratify UNCLOS despite being in favour of all other aspects of the Convention. In mid-2008 the United States changed its position, though at the time of writing (January 2009) it has still not ratified the Convention. However, it is anticipated that the new US President, Barack Obama, will ensure ratification, having indicated that intention during the Presidential campaign. That said, US oil interests hope the United States will never ratify, believing that the ISA is intent on blocking development of the seabed or on imposing taxes which will deter its development. US oil interests claim that ISA's position is linked to Russian attempts to sequester the Arctic seabed, which a US coalition (Arctic Oil and Gas, and its associates) claimed on 9 May 2006. The claim comprised the 'Arctic Ocean Commons' a vast area of the Arctic Ocean (chiefly the area below the permanent (i.e. summer) sea ice). The claim has not been recognized by any international organization. US oil interests also claim that the ISA's position, and the resulting lack of development, will bring misery to the world's poor, though it hardly requires an overwhelming cynicism to believe the main concern is profit rather than altruism.

The North-West Passage

The North-West and North-East Passages represent particular problems in terms of the Law of the Sea, though the way in which the problems have been dealt with differ markedly.

As mentioned above, the 1988 Agreement on Arctic Cooperation between Canada and the United States sought to advance the 'shared interests in Arctic development and security' of the two nations. Ironically, the existence of the Agreement has not stopped the two nations having a very marked difference of opinion over the status of the North-West Passage, despite navigation in the Passage being (the Canadians claim) crucial to the shared security of the two nations, and of keen interest to both in terms of future Arctic development.

The 1988 Agreement was important not only for what it included, but what it excluded and the extent to which the exclusions were deliberate. In 1969 Humble Oil & Refining Co., in an effort to prove that the North-West Passage was a viable route (as a route between Europe and Asia it is several thousand kilometres shorter than the Panama Canal route) decided to attempt a transit. The company modified the oil tanker *Manhattan*, replacing the bow with a longer, heavier section, modifications which effectively turned the ship into an ice-breaker. The work was accomplished in Massachusetts, and the ship then sailed west through the Passage, accompanied by the US Coastguard ship *Northwind* and the Canadian Coastguard ship *John A. Macdonald*. The ship reached Prudhoe Bay where a ceremonial barrel of oil was taken on board. The ship then returned through the Passage, again accompanied by Coastguard vessels.

There had been previous transits of the Passage by US vessels – by the Coastguard ships *Storis*, *Spar* and *Bramble* in 1957, and by two nuclear submarines, the *Seadragon* in 1960 and the *Skate* in 1962, the latter making a submerged transit. But despite these earlier voyages, and the fact that there was a Canadian government official on board the *Manhattan*, the latter's transits caused a public furore in Canada. There is even a story that during one of the transits the *Manhattan* was stopped by an Inuit group who demanded, received and granted a request for right of passage. If the story is true, the confrontation must have been a marvellous spectacle, a handful of native hunters against a 1,005ft (306m) long ship with a cargo capacity in excess of 106,000 tons.

The trips in 1957 and 1962 had been viewed as aiding Canadian security: in 1957 the 1939–45 war was a recent memory; 1962 was the year of the Cuban missile Crisis. In 1969 the world seemed less hostile and, of course, the *Manhattan* voyages were a commercial, not a naval, venture. But though this certainly stoked the controversy, the real issue was the revelation that Canada had exercised no more than the claim of 3 nautical miles of territorial sea around any of the islands of its Arctic archipelago, something unknown to the great majority of the population. This lack of a claim meant that the *Manhattan* had, therefore, in general sailed the high seas, a fact which both surprised and angered the Canadian public. In 1970 the owners of the tankers announced that they would attempt a further transit, causing renewed public anxiety.

Forced to act, the Canadian government extended its territorial waters to 12 nautical miles, both from the mainland and from the shores of all the islands of the Arctic archipelago. There are several potential 'North-West Passages', but the new limit placed almost the entire length of all of them in Canadian waters. The government also enacted the Arctic Waters Pollution Prevention Act (AWPPA) which gave it jurisdiction over 100 nautical miles of water measured from the low-water mark of the mainland and all islands, and the right to enforce standards of vessel construction and operation to any ship sailing in those waters. Within the Act 'Arctic waters' were defined as being in either the liquid or solid state. The deposit of 'waste' in such waters was prohibited, with 'waste' being defined as anything which could degrade the waters to the detriment of water usage 'by man, or by any animal, fish or plant that is useful to man.' Though at first glance this seems a curious, homo-centred definition, in reality anything in Arctic waters can be said to be useful in one way or another.

The Act meant that Canada could veto a transit by any vessel which failed to meet the standards it set. This was actually a very clever ploy by the Canadians: the *Manhattan* had sustained relatively serious damage during its voyages, damage that could possibly have resulted in a loss of oil had it been loaded. By transferring its concerns to environmental protection, the Canadian government had made no overtly sovereignty-based claim to the North-West Passage, merely acting to protect the fragile Arctic waters surrounding the Arctic archipelago over which it did claim sovereignty.

After lobbying other Arctic nations, and obtaining the support of Norway, Sweden and, crucially, the USSR, Canada went to the United Nations where UNCLOS was already being discussed. As a result of Canadian pressure, when UNCLOS was finally adopted in October 1982, it included, as Article 234:

> Coastal States have the right to adopt and enforce non-discriminatory laws and regulations for the prevention, reduction and control of marine pollution from vessels in ice-covered areas within the limits of the exclusive economic zone, where particularly severe climatic conditions and the presence of ice covering such areas for most of the year create obstructions or exceptional hazards to navigation, and pollution of the marine environment could cause major harm to or irreversible disturbance of the ecological balance. Such laws and regulations shall have due regard to navigation and the protection and preservation of the marine environment based on the best available scientific evidence.

At the time, embryonic concerns over global warming had not led to suggestions that a North-West Passage would be ice-free any time soon. Now that seems likely to be a reality in decades (or even less), the protection offered to the Canadians by Article 234 will come into question. However, UNCLOS also includes (Article 192 of Part XII) the provision that 'States have the obligation to protect and preserve the marine environment' which Canada could interpret as giving it the right to legislate to provide environmental protection to ice-free water between the islands of its Arctic archipelago.

The 1970 *Manhattan* voyage did not take place, but the US government pointed out that this did not alter its view that, despite the Canadian legislation, formal permission for a transit of the Passage was not required. The US position was that the Canadian imposition of a 12-nautical mile limit to its territorial waters was irrelevant as the Passage was actually a channel (or a series of channels) connecting waters which were 'high seas'. Therefore, the United States claimed, the Passage was itself 'high sea'. The US argument rested on the Corfu Channel Case heard by the International Court of Justice in 1949 (the first case to have been heard by the Court). The case had been between the UK and Albania and arose from an incident on 22 October 1946 in the narrow channel between Corfu and Albania when two Royal Navy vessels struck mines, causing both severe damage to the ships and loss of life. The argument before the Court related chiefly to responsibility for the incident and compensation due to the UK if Albania was found responsible (Albania was found responsible and ordered to pay almost £1 million in compensation), but was also asked to decide whether the vessels had the right to pass through the channel in peacetime, i.e. had the British ships violated Albanian sovereignty? The Court's decision on the latter question was that as the channel linked sections of international waters, transit was acceptable without prior agreement and that the channel should be kept open and free of hazards.

Taking its cue from the Corfu case, the United States maintained – and still maintains – that as transits of the North-West Passage use channels between the islands of Canada's Arctic archipelago, and these channels link areas of international waters (i.e. 'high sea') – essentially the Pacific and Atlantic oceans – the Passage itself is international water. For their part, the Canadians question the validity of the Corfu Channel case in respect of the Passage, believing that the waters of the Passage represent 'historic internal waters' and, as such, fall outside the jurisdiction of the Court. They also note that the position with respect to the case is not as clear cut as the United States would wish it to be, as the International Court noted that not only geography, but function was an issue. As well as defining the Corfu Channel as international water because of its geographical position, the Court noted that the channel received a substantial amount of traffic. The Canadians have seized upon this rider to suggest that the fact that the North-West Passage has had, and continues to have, very few transits means it fails the functionality test and cannot be regarded as international regardless of its geography (or their claim of historic internal waters). The suggestion that climate change might in the future mean that the Passage will receive increasing amounts of traffic does not constitute a case now. Canada also noted, and continues to maintain, that the Passage is not a single channel, but a series of connected channels and the Corfu Case is not, therefore, fully representative of the archipelago.

In 1985 the US Coastguard ice-breaker *Polar Sea* made a transit of the Passage. There were Canadian coastguard officers on the ship, but neither the US nor the Canadian government made any claim that their presence amounted to the formal granting of permission for the voyage. Despite public opinion and several attempts to force the Canadian government to issue an official protest over the voyage they did not do so. However, they did issue a statement stating that the voyage did not 'compromise in any way the sovereignty of Canada' over its northern waters, nor did it 'affect the quite legitimate differences of views that exist between Canada and the United States on that question.' Nevertheless, the voyage did result in a statement setting out the Canadian position in a definitive way. This included the drawing of straight line boundaries at the western and eastern ends of the archipelago, with a claim to sovereignty over all land and waters within them, and the extension of Canadian civil and criminal law to the area enclosed by these new boundaries. The boundaries recognized that

> Canada's sovereignty in the Arctic is indivisible. It embraces land, sea and ice. It extends without interruption to the seaward-facing coasts of the Arctic islands. These islands are joined, and not divided, by the waters between them. They are bridged for most of the year by ice. From time immemorial Canada's Inuit people have used and occupied the ice as they have used and occupied the land.

To back its claim the Canadians increased air and maritime activity in the area and placed a contract to build a new ice-breaker, a contract which was later cancelled to the dismay of many.

But despite the redrawing of boundaries and the claim to overall sovereignty, the position of the North-West Passage remains controversial, with both the United States and the European Union continuing to claim the Passage as international waters. Recent Canadian attempts to bring the United States around to its position have been based on the idea that supporting Canada's position would enhance US security. The Canadian argument is that a free-for-all in the Passage would give potential enemies and terrorists the ability to use the seaway as a backdoor to the United States. At the very least, Canada seeks US support for the AWPPA Act, which would deter aggressive users of the Passage as they would know that entering Passage waters could result in an inspection of their vessel. These overtures have yet to draw a positive response from the United States, the official position remaining that the two countries (and, by implication, all those countries which side with the US view) agree to disagree on the legal position of the Passage.

There are initiatives as well, some involving suggested joint agreements which bother many Canadians as they feel that attempts to break the deadlock over the Passage's position would necessarily involve a weakening of Canadian claims of sovereignty. The problem faced by the Canadians is that policing the Passage at a time when its waters are becoming increasingly easy to navigate is a massive task, but if the waters are not policed then inevitably unregulated traffic will increase. The Canadians are therefore working on two fronts.

In late 2007 the RADARSAT-2 satellite was successfully launched by the Canadian Space Agency using a Russian Soyuz rocket. After a period of in-space commissioning the satellite began commercial operation in May 2008. The satellite is claimed to have 100m positional accuracy and a resolution of 1–3m and can be used for the monitoring of both ice and shipping. To the astonishment of many, it was planned to sell the satellite to a US company, but its usefulness in patrolling Arctic waters caused the sale to be blocked by the government. In addition to the satellite, Canada is also intending to commission a land-based detection system which can use both radar and laser-based systems to monitor shipping in Lancaster Sound. Although in principle a ship could use the Hudson Strait, Foxe Basin, Fury and Hecla Strait and Bellot Strait to reach Victoria Strait, or could use Smith Sound and Nares Strait to round the north end of Ellesmere Island, these alternatives involve narrow channels which are currently ice-choked more or less continuously and will be hazardous even if completely ice-free (which the Ellesmere route will not be for many years even according to the most pessimistic assumptions). The only realistic route is therefore through Lancaster Sound, the (relatively) narrow channel between Baffin and Devon islands. A land-based system on northern Baffin Island could therefore monitor Passage shipping.

In addition to these monitoring systems, Canada has also decided to build a fleet of Arctic Offshore Patrol ships, though the ice-breaking capabilities of these has been questioned by some experts, leading to the unkind suggestion that 'Polar Class' vessels are little more than 'slush-breakers'. In 2006 aircraft of the Canadian Air Force also flew the first Northern Sovereignty Patrol, from Inuvik to Cambridge Bay. While Arctic flights have been routine as part of the joint US-Canadian North American Aerospace Defense Command (NORAD), this was the first flight specifically to reinforce Canadian Arctic sovereignty claims. Such flights are also likely to become routine in the future, probably directed, in part, from the Forward Operating Base established at Inuvik airport. The Canadians are also planning to open an army base at Resolute Bay on Cornwallis Island.

Yet despite these positive steps, which have been welcomed by many who see a need for Canada to be much more assertive of its Arctic claims in the future, there are still critics, one of the major areas of criticism being the continued optional nature of the registration of ships on NORDREG, the Canadian Arctic marine traffic system. Most of the current traffic in the North-West Passage at present is tourist-based, and although these, and other, vessels are encouraged to register (and registration is free), it is not compulsory to do so. Most ships do register as registration allows them access to up-to-the-minute ice information from the Canadians. However, with ice coverage shrinking, and ice information available on the internet, some ships do not register. In both 2007 and 2008 Arctic cruise ships failed to register.

The North-East Passage

Some time in the sixteenth century, the discovery of the horn of a unicorn on the shores of the Kara Sea caused great excitement in Europe. It is not clear at what point those who were discovering unicorn horns found out they were the tusks of the narwhal, but it is likely that, even after they made the connection, the sale of unicorn horns was so profitable that the information was kept secret from potential buyers. What is certain is that the first voyages to discover an alternative route to Cathay headed north-east. This is hardly surprising, the merchants sponsoring the journeys 'knowing that Unycorns are bredde in the landes of Cathaye, Chynayne and other Oriental Regions.' If unicorn horns had been washed up by the Kara Sea, then a sea route eastward clearly existed. Early attempts at finding a North-East Passage were by the Dutch and the British, Richard Chancellor having reached Kholmogori (which later became Arkhangelsk – Archangel) in 1553 and Willem Barents reaching the Kara Sea in 1594. But the ice of the Passage, unpredictable and deadly, soon called a halt to the ventures in favour of the apparently more promising idea of heading north-west.

In the wake of the conquest of Siberia, Russians explored the northern mainland coast, Dezhnev sailing through the Bering Strait in 1648 and the explorations of the Great Northern Expedition successfully mapping the northern mainland coast. That clearly implies that the Russians knew a great deal about their northern coast, and that is indeed the case. As early as 1619 the Tsar ordered that anyone disclosing information on the route from the Kara Sea to the Ob River would be executed, so important was it for the fur trade and, consequently, the Tsar's coffers.

The Russians had gained a more or less complete understanding of the coast by the middle of the eighteenth century and Mikhail Lomonosov – after whom a significant undersea ridge in the Arctic basin is named – suggested a transit of the North-East Passage. This was agreed, but Lomonosov died (in 1765) before the voyage could take place. It went ahead under the command of Vasili Chichagov, but rather than attempting to follow the coast, the Russians headed north to 80°N before turning east: not surprisingly they were soon stopped by impenetrable ice. The failure stopped any further attempt, the difficulties and unpredictability of the ice acting against the route's usefulness for trade, and it was not until 1878–9 that the first transit was made, west–east, by the Swede Nils A.E. Nordenskiöld. In the early 1900s, ice-breakers sailed through the Passage, with the first single-year transit being made in 1932 by the ice-breaker *Sibiryakov*. In the same year the USSR established the Northern Sea Route, a shipping lane, which, they hoped, would cut journey times – the route cuts the distance between the Russian Atlantic and the Russian Pacific by half – and so makes trade cheaper. Early journeys held out significant promise, but the loss of the *Chelyuskin* in 1933–4 (even though it involved the loss of only one life and that due to an unfortunate accident rather than as a direct result of the ship's entrapment) suggested that the route would always be hazardous.

The position of the North-East Passage is both similar to, and different from, that of its Nearctic counterpart. With the present distribution of summer sea ice, there is only one Passage, a vessel needing to head between the southern edge of the four archipelagos and Wrangel Island, and the Russian mainland. Russia has a claim to the seas traversed being historic waters. It also has a claim to their being internal waters as in 1985 the USSR drew straight lines from the Arctic islands to the mainland and declared all the waters within the lines to be internal. In doing so the USSR was making the same claim for the waters between its northern shore and its Arctic archipelago as Canada was to do for the waters of the North-West Passage. However, the USSR did not extend the claim of internal waters to lines drawn from Franz Josef Land, which is a considerable distance from the mainland. On the other hand, the definition of a border differs in Russia from that of Western nations. In the latter it is a line, while in Russia it is a zone, a ribbon of variable width. Russia's border zone is also subject to travel restrictions. In Britain, anyone wishing to sail to Norway, as an example, needs no more than a passport to start out and the usual port of entry regulations when arriving. In Russia a permit is required to visit the border zone and exit from Russia may only be from designated points of exit. The idea of a border zone dates from Soviet times when the country needed to be protected from both outside and within, the latter requiring that anyone wanting to visit the border area had to be screened and issued with approved documentation. These regulations were abolished when the USSR broke up, but were re-instated in 2002.

Though the drawing of the eastern and western bounding lines, and the claims of historic/internal waters make the positions of the two Passages similar, there is a critical difference. While US decisions to sail through the North-West Passage have been accepted by Canada (with due allowance for the fact that the

Canadian government does not agree that the voyages constituted any threat to their position on the waters of the Passage), there have been no similar voyages through the North-East Passage. In the early 1960s several US scientific expeditions were made by both naval and coastguard vessels into Arctic waters north of the USSR. It is clear that the Soviet authorities were not happy with these trips, though the level of protest was muted. However, when in 1965 the United States announced that the coastguard vessel *Northwind* was to travel between the Kara and Laptev seas by way of the Vilkitski Strait between Bolshevik Island (the southern island of Severnaya Zemlya) and the Taimyr Peninsula, there was an immediate, and strong, Soviet reaction, and the plan was abandoned. Two years later the Americans tried again with two different coastguard vessels, the *Edisto* and *Eastwind*, but again abandoned the idea in the face of strong Soviet protests. In each case the Soviet protest was based on the fact that both the channels through which the vessels might pass (there are northern and southern channels each side of the Geyberg Islands at the western end of the Strait) are less than 24 nautical miles wide and as such are USSR territory. The voyages would therefore, the Soviet government claimed, be a violation of USSR borders and could only occur if permission was requested and granted. Given the relations between the countries during this particular period of the Cold War, it is not surprising that the United States chose neither to seek permission, as that would have granted legitimacy to Soviet jurisdiction over a channel the United States believed was international water, nor to press ahead with the voyages even if they felt their legal position was sound.

Since the 1939–45 war, the Soviet Union and now Russia has maintained the Northern Sea Route, particularly after the exploitation of the mineral resources of northern Siberia required the establishment of new ports. A fleet of Russian ice-breakers, aided by aerial reconnaissance and by radio weather stations, keeps the route navigable from June to October. Shortly after the US decision to abandon the 1967 voyage, the Suez Canal was closed during the Six-Day War. The Soviet government offered foreign companies the alternative route through the Passage, suggesting it would provide ice-breaker support. That offer was not taken up, and the response was equally negative after the similar offer made in Gorbachev's Murmansk speech, where he stated 'I think that, depending on how the normalization of international relations goes, we could open the Northern Sea Route to foreign ships under our ice-breaker escort.'

At present, while tourist vessels, often Soviet or Russian built and owned, regularly ply the waters of the North-West Passage, often putting passengers ashore on islands or the northern mainland coast, there are very few such trips to the North-East Passage or to the islands of Russia's Arctic archipelagos. It is not difficult to understand why: air travel to and within the western Arctic is both easier and, to the casual observer, more reliable, than in Russia and Alaska, Canada and Greenland are much more tourist-friendly than Russia, where much of the bureaucracy and downright suspicion of the Soviet era lingers on. Add to this the fact that Russian officialdom is inherently hostile to outsiders visiting remote areas of the country, something which even limits the possibilities of Arctic travel by Russian-based tourist companies, and it is easier to understand why Franz Josef Land has very few, sometimes no, visitors in a season while nearby Svalbard throngs with them. There is another reason, and that is reflected in the difference in the positions of Canada and Russia over their respective claims for sovereignty over the waters north of their coasts. Simply put, the Canadians are concerned with their position in the world, a position which requires an acceptance of the views of other nations, and a political conscience which recognizes the status of international law even if that recognition means doing things, or agreeing to things, which are, if not exactly detrimental to Canadian interests, at least not helpful to them. The Russian position is much more difficult to define, since it consists of two conflicting streamlines: the official position of the government, which is usually hostile to its own people, and position of Russian people themselves, about which they are very rarely asked. The 'official Russian' position in dealings with the international community is along the lines that if the rest of the world does not like what the Russians are doing or proposing, then the rest of the world must do the other thing – Russia will act strictly according to 'Russian official' interests. The problem is exacerbated by the present nature of Russian bureaucracy. If several companies attempted to set themselves up to aid tourism to, for instance, Franz Josef Land, then Moscow bureaucrats would veto the attempt for two reasons: firstly that would be too many companies to control (whether the number was 2 or 3 or 100); secondly because the money earned would not be processed through the Moscow bureaucracy and too much might therefore reach ordinary Russians rather than being creamed off by the bureaucrats themselves. The best solution is, therefore, one company which can be controlled and milked, or no companies at all.

As the US position had become increasingly similar under the Bush administration, the Arctic scene was set for conflict, at least in the diplomatic sense. It will be interesting to see what changes in policy the more environmentally-friendly Barack Obama makes as President and whether the necessity to unpick the appalling changes in environmental law ushered in by Bush during the last days of his presidency, in an effort to negate Obama's enthusiasms, will mean the US taking its eye off the Arctic ball.

A section of the Trans-Alaskan pipeline which takes oil from the Prudhoe Bay fields to Valdez.

Chapter Five
Exploitation: The Second Wave

First the White Man took all the beaver, then he came back and took all the trees. Now he is taking the rocks.
Canadian First Nation man watching the activities of mining company operatives on his ancestral homeland

The fur trappers and whalers who travelled to the Arctic had no concept of conservation; their objective was to kill as many animals as they could. They were men engaged in a risky business, one in which they gambled not only their futures, but their lives. American whalers were not paid a salary for their efforts, but took a share of the profit of the voyage. If few whales were taken, the crew, who paid for the clothing and other essentials they obtained from the ship's store, might end up owing the ship's owner money at the end of the voyage, a system which encouraged foolhardy bravery and a pitiless, single-minded pursuit of the quarry. With the first wave of exploitative incomers, those who sought Arctic animals, gone, it might be expected that in a more enlightened age those who followed them might have more respect for the land. That was not to be the case.

In the immediate post-1939–45 war period, apart from in Russia, the Arctic was not seen as a valuable asset, a region rich in mineral resources. It was, rather, a place where the main exploitation had been completed, any possible mineral wealth being entombed in an icy shroud that made development difficult and expensive for machinery, and unpleasant and potentially lethal for people. But the Arctic had other uses. After 1945 the Western world enthusiastically began to explode nuclear devices on the atolls of the Pacific and in the dry lands of Nevada and Australia. As a consequence, when an increasingly isolated Soviet Union, seeking to maintain parity in the nuclear weapon field, looked for a desolate place with a low population density for its own nuclear testing, it turned its eyes to the Arctic.

In 1955, having forcibly removed the 536 residents (Nenets and Russian fur trappers) from the southern island of Novaya Zemlya, the Soviets carried out a nuclear bomb test under the sea of Chernaya Bay on the western side of the island. In removing the locals, the Soviets were actually only learning from the Western powers who had forcibly removed native populations (eg. the US removal of the Micronesian population from Bikini Atoll in 1946), and in using the Arctic they were also avoiding the problems caused by people wandering into test areas, as the Japanese *Lucky Dragon* fishing boat had to the Bikini test area in 1957. With no local population and no possible intruders, the Soviets avoided the possibility of later legal cases. They were also more careful about allowing the native population to return, again learning from the US mistake in letting the original inhabitants return to Bikini Atoll in 1968 when the radiation levels were too high. In that sense, the Soviet decision to use the Arctic was justified. Of the 2037 nuclear tests carried out during the period between 1945 to 1998 a total of 212 were made in Arctic. The USSR dominated nuclear testing in the Arctic, with 209 tests at two sites on southern Novaya Zemlya and a third site on the northern island, the other three being US tests on Amchitka Island in the Aleutians. The Soviet tests included the largest bomb ever exploded, a 58 megatonne hydrogen bomb in October 1961. Many of the tests were atmospheric, resulting in the uncontained spread of radioactive material. The official Russian position is that while activity levels at the three test sites are high, the level on the remaining island is 'only marginally above background.' Public visits to the sites are banned, but interestingly the quoted levels of radioactive contamination are lower than those quoted for Bikini Atoll, where diving is now permitted.

Although US nuclear bomb testing took place chiefly in the Nevada desert, three tests were carried out on Amchitka Island in the Aleutians (in 1965, 1969 and 1971) despite the island being part of the Aleutian Islands Wildlife Refuge created in 1913. As everywhere, military requirements supersede all others. There were also plans to excavate a harbour at Ogotoruk Creek, near Cape Thompson about 200km (124 miles) south-east of Kotzebue in Alaska with one or more nuclear devices. As part of the planning process for the scheme (known as Project Chariot) trace radioactive material (Cs137 and Sr85) was placed in locations around the site. The position of this material was not on the public record and came to light only in 1992, five years after use of the site had been granted to a local man, as a result of an independent search of declassified documents. The material was then removed. Radioactive traces of the US Nevada and Pacific atmospheric tests, moved worldwide by the jet stream, are detectable in Alaska and across Canada. Researchers actually use them to date permafrost lake sediments.

At the same time as the tests were being carried out, the USSR was dumping an estimated 11,000 containers of highly radioactive material in the Barents and

Kara seas. The nuclear debris on the seabed also apparently includes the reactor units from the nuclear ice-breaker *Lenin* which, Western sources maintain, had at least two accidents in the 1960s, one of which resulted in a core meltdown and the jettisoning of the core and associated debris. In April 1989 the nuclear submarine *Komsomolets* sank about 120km (75 miles) south-west of Bear Island following an on-board fire. The submarine's reactor and, Western sources maintain, two nuclear-tipped missiles or torpedoes, went down with the boat. More recently, the nuclear submarine *Kursk* suffered a catastrophic explosion (in the bow torpedo compartment, away from the reactor) and sank in the Barents Sea. The attempt to rescue survivors trapped in horrific conditions involved several nations, but failed; all 118 men on board died. The bow of the submarine was cut free and blown up *in situ*; the rest of the boat, including the intact reactor, was salvaged. Aged, redundant nuclear submarines have also been left to decay in bays along the northern coast of the Kola Peninsula. All nations with nuclear submarines regularly operate them beneath the sea ice of the Arctic Ocean as it provides a useful cover against the prying eyes of spy satellites.

During the 1939–45 war the United States created seventeen military sites on Greenland. Then, during the Cold War, several stations of the US Distant Early Warning (DEW) system were established on the island; most have now been abandoned, leaving behind spilled oil and toxic waste that have been implicated in higher cancer rates of villagers close to some sites. None of the sites held nuclear material, though in contravention of an agreement with the Danish Government the United States were storing nuclear weapons on Greenland, at least until the 1960s. In 1951, after the forced resettlement of native peoples, the Americans built the Thule airbase in north-west Greenland. In 1968 a B52 bomber crashed into the sea 12km (7½ miles) west of the base (though some reports have the aircraft crashing on to the ice and catching fire). Although it was not at first admitted, the bomber was carrying four hydrogen bombs. The nuclear fusion of a hydrogen bomb is triggered by a fission device: in the case of the four bombs on the B52 this was plutonium based. The fusion elements of the device present minimal risk to the environment, but plutonium is both dangerously radioactive and chemically poisonous. Despite attempts to maintain secrecy over the incident, the Americans have admitted that not all the plutonium was recovered, with an amount variously stated to be 500g–1.8kg (18oz–4lb) having gone missing. However, there are credible, but unconfirmed, reports that one bomb, with its fission trigger, was not recovered and had probably reached the seabed. The plutonium loss could then be up to 12kg (26lbs). There are reports from the clean-up after the event that suggest poor environmental control, with radioactive material leaching into waterways. Some workers unsuccessfully sued the US government for compensation. Inuit hunters have also claimed deformities in local muskoxen and seals. Their attempts to have the base closed have so far failed.

Peaceful uses of nuclear energy have also reached the Arctic. Russia has four nuclear reactors at the Polyarnye power station, near the city of Polyarnye Zori close to Murmansk. The reactors are VVER-type, not the RBMK design involved in the accident at Chernobyl. However, the two older reactors at the site are considered by Western nuclear experts to be as 'dangerous' as the Chernobyl type, though use of the emotive term 'dangerous' is seen by many as politically motivated. The reactors provide power to local heavy industry at Monchegorsk as well as local towns; some electricity is also sold to Finland. There is also a nuclear power station at Bilibino in Chukotka, built to power local gold and tin mines. The Russian plants are the world's only nuclear power stations built within the permafrost zone, though they were constructed on rock outcrops.

One further, bizarre, episode involved the operation of a nuclear reactor (a type PM-2A reactor, generating 1.5MW of electrical power as well as providing steam for camp heating) at Camp Century (at 77°11′N, 61°08′W, about 225km (140 miles) east of the Thule air base) beneath the Greenland ice cap, as part of Operation Iceworm, a study to see if it was possible to construct nuclear missile sites beneath the ice. The reactor's major containment was the surrounding ice. Although the presence of the reactor was made public in 1960, the year it began operation, it took until 1963 for Danish authorities to protest sufficiently for the reactor, and the camp, to be dismantled and moved back to the United States.

Both Finland and Sweden also have nuclear power stations, though these are situated in the south of each country.

Russia also has three nuclear fuel reprocessing plants: at Krasnoyarsk, Mayak and Tomsk, each far removed from the Arctic. The first discharges into the Yenisey River, the latter two into the Ob. However, as a result of the Gulf Stream, the major source of radioactive particles in the European Arctic is the UK Sellafield site, the Stream sweeping the particles along the Norwegian coast and on towards Svalbard. In 2002 AMAP reported that

> In general, levels of radionuclides in the Arctic are declining. The exceptions are seawater concentrations of the long-lived water-soluble fission products 99Tc and 129I. This is due to increased releases from nuclear fuel reprocessing in Western Europe and the releases of radioactive technetium to the marine environment from

> Sellafield. There is evidence that sediments are now a source of Pu and 137Cs to the Arctic. Previous releases, such as those from Sellafield that have deposited in Irish Sea sediments, are being remobilized such that these deposits now act as sources to the Arctic.

Although the Chernobyl nuclear power plant was remote from the Arctic, the accident in 1986 caused Caesium contamination, which led to the slaughter and burial of 30,000 reindeer in Scandinavia. Lichen in areas affected by radioactive material is likely to remain toxic for at least another decade.

By the 1960s, the Arctic's potential as a source of minerals was beginning to be realized by all Arctic nations, though it would be wrong to say that the second wave of Arctic exploitation dates from that time. The exception was the Soviet Union where many existing mines in the Arctic were being closed (for instance on the Taimyr and Kola Peninsulas), mainly due to the dismantling of the gulag system. One older industry is coal mining, carried out on both Svalbard and Bear Island by the Norwegians, and by the Soviets (and now the Russians) on Svalbard. Although there had actually been some small-scale extraction earlier, coal mining on Svalbard began in earnest in 1906 when two Americans, Frederick Ayer and John Longyear, formed the Arctic Coal Company to mine deposits in Adventdalen (above the archipelago's 'capital') they had purchased from a Trondheim company. In 1916 the Americans' company was taken over by the Store Norske Spitsbergen Kulkompani A/S (SNSK), formed by Norwegian businessmen. SNSK and its operating companies still mine at Longyearbyen, though the main production is now at Svea, a mining community about 60km (36 miles) south-east of the capital which had originally ceased production in the 1960s. The Svea mine is estimated to be capable of yielding 32 million tons of coal. Mining on Svalbard is controlled by a Code prepared and administered by the Norwegian Government. Claims made by SNSK under this Code are estimated to cover around 100 million tons of coal deposits. The Longyearbyen mine now produces coal only for the power station which supports the town. A third Norwegian mine, at Ny Ålesund, operated by the Kings Bay Coal Company, was closed in 1963 after an explosion in 1962 which killed twenty-one men. Ny Ålesund is now a scientific research facility. On Bear Island the Norwegians extracted 116,000 tons of coal between 1916 and 1925.

Under the terms of the Svalbard Treaty, the USSR began coal mining on Svalbard in 1932. The Soviets opened and operated three mines, Barentsburg, Grumant and Pyramiden. Grumant closed in 1962 and Pyramiden in 1998, but at Barentsburg, Trust Arktikogul produces about 120,000 tons annually. The known deposits are expected to allow continued extraction until 2020. Recently the Russians stated their intention of re-opening the Grumant mine, which lies 25km (15½ miles) north-east of Barentsburg, which, they claimed could produce up to 500,000 tons annually for 50–60 years. The Russians also stated that they were considering reopening the Pyramiden mine, which lies at the head of Billefjorden, a north-eastern extension of Isfjorden, where, they claimed, deposits of at least one million tons remained.

While coal extraction has been, to date, the most consistent mining activity in the Arctic, the most famous has been the extraction of gold, and the Klondike and Alaskan gold rushes have become the stuff of legend. The Klondike strike was in 1896 near Rabbit Creek (later renamed Bonanza Creek), a tributary of the Klondike River, close to the latter's confluence with the Yukon River. The strike resulted in an estimated 40,000 hopeful men converging on Dawson City, a town that grew from nothing to accommodate them and the services provided to them. The romantic vision of the gold rush centres around stern-wheeler ships plying the Yukon River, but the truer image is of men struggling to climb the Chilkoot Trail in the biting cold of a Yukon winter and of few making enough money to warrant the effort and time put in. Gold was first discovered in south-east Alaska, near Juneau, but the richer and more famous find was at Nome in 1898. The gold rush that followed brought miners, and entrepreneurs who sold articles and services to the miners, as well as finding numerous other ways of relieving them of their cash. By 1900 Nome was Alaska's biggest town, its inhabitants including the legendary gambler and lawman Wyatt Earp who arrived to assist in the removal of money from lucky miners. Today gold is still extracted at Dawson City, though the pay gravel dredgers at Nome have been still for many years. Gold is also extracted near Fairbanks in Alaska, while in Russia gold mines in Siberia account for almost 70 per cent of the country's production. The richest mines in the Siberian Arctic were in Chukotka, but there are also mines along the Kolyma river and in Kamchatka, with deposits in the northern Urals also claimed to be rich. Gold has also recently been worked at Nalunaq, near Nanortalik in southern Greenland.

Other early mineral extractions which have ceased and which are unlikely to be become economic in the future include the extraction of galena, a lead ore, on Bear Island during 1925–30 and the extraction of cryolite at Ivittuut, near Narsaq in south-western Greenland. Cryolite (sodium hexafluoroaluminate) was an early source of aluminium, and was later used as a flux in the production of aluminium from bauxite. So important was the production of the mineral during the 1939–45 war that the United States effectively annexed the mine to support its war effort. During the 1960s production slowed, in part because other materials had been developed which aided metal extraction from bauxite. The Ivittuut cryolite deposits were exhausted in 1987 and the town/mine were abandoned.

The smelting of iron from iron ore dates back 2,500 years in Arctic Sweden, though industrial production only began in the seventeenth century. The extraction of the magnetite and haematite ores in the twentieth century led to the development of the Swedish car industry. (Both Saab and Volvo are now owned by international conglomerates, General Motors and Ford respectively. Saab sold only its car business and continues to manufacture planes, aerospace components and other products; Volvo also continues to manufacture trucks and buses.) At present LKAB (Luossavaara-Kiirunavaara Aktiebolag) operates mines at Kiruna and Malmberget (the name translates as 'iron mountain'), which have proven reserves of 3,000 million tons. The Kiruna mine is the world's largest underground iron ore mine, producing about 15 million tons of ore products annually, the ore having a 60 per cent iron content. Although once a very significant iron ore producer, Arctic Sweden's production is now less than 2 per cent of world annual production. For comparison, China produces about 33 per cent of world production, with Brazil and Australia producing about 17 per cent each. Northern Sweden also has the largest lead ore deposits in Europe, at Skellefteå, and significant deposits of copper. Silver was also once mined at Nasafjall. Other metal mines include a cadmium, lead, silver and zinc mine at Nanisivik at the northern end of Baffin Island, which closed in 2002, and operating lead and zinc mines on Canada's Cornwallis Island. Close to Uummannaq on Greenland's west coast lead and zinc were mined until the 1990s. There was also a marble quarry in the area which operated successfully for many years before ceasing production in 1972.

Of presently operating metal mines the most important are Red Dog, about 150km (90 miles) north of Kotzebue in north-west Alaska, and the mines in Russia. The Red Dog Mine is located on land owned by the NANA Regional Corporation, one of the thirteen Alaska Native Regional Corporations created under the Alaska Native Claims Settlement Act of 1971 (see pages 78–9). It is operated by Teck Cominco, a Canadian mining company. The mine is the world's largest producer of zinc, and has the largest known zinc reserves. Annual production of zinc exceeds 500,000 tons, in addition to which over 120,000 tons of lead and 75 tons of silver are also produced annually. Concentrated ore is trucked along an 85km (53-mile) haul road (which passes through the northern edge of the Cape Krusenstern National Monument, an area of wetland important to Arctic birds, and of raised beaches which are important archaeologically), to a seasonal port on the Chukchi Sea where it is stored until thawing sea ice allows shipping south: shipping is currently possible for about 100 days annually. The mine is only accessible by air, as is the port during the 250 days when the sea is frozen. At present Teck Caminco has permission to mine until 2012 and has applied for permission to expand its operations to encompass known reserves which would allow mining until the 2030s.

Mining at Red Dog has been responsible for interesting conservation issues. Before Teck Camino began operations the local streams were too toxic to sustain life, as water percolated through the metal deposits becoming contaminated with heavy metal, not only zinc and lead, but cadmium, aluminium and other metals which are found in traces in the metal ore deposits. Stream water also had a low pH. Ore extraction and waste treatment has meant a reduction in water toxicity and Arctic grayling are now found in streams formerly devoid of life. However, studies have revealed the presence of heavy metals in plant life beside the haul road to the coast as a result of dust release during transport and ore spills, and in the longer term leaching of heavy metals into local water will remain an issue. The US Environmental Protection Agency (EPA) has declared Red Dog to be the producer of the most toxic waste of any operation in the country, but that view has to be tempered with the understanding that 'toxic waste' in the definition used by the EPA includes waste rock from ore extraction if the rock contains greater than 2 per cent sulphide concentration. As waste rock constitutes some 99 per cent of Red Dog's waste, and that concentration of sulphide occurs naturally, it can be argued that the waste should be excluded from the EPA's calculation. That argument is one which exercises both sides of the conservation debate. Of more concern are the fundamental issues of whether a huge open-cast mine should be allowed in an area of pristine wilderness, and the worries of local native peoples. Their concerns centre on the effect the haul road had on animal migration routes, the possible effect on local sea mammal populations of a proposed extension of the mine's port facilities, and the effect of long-term toxic leaching on the local ecosystem.

Raglan Mine, in Nunavik, northern Québec, is a nickel mine set on what is one of the largest nickel deposits in the world. Mining began in 1997 and is expected to continue until the 2030s. Ore is extracted from three underground mines and an open cast working. It is concentrated on site, then trucked 100km to a seasonal port on Deception Bay from where it is shipped to Québec City, and then rail transported to Sudbury, Ontario for smelting. Although nickel is the prime output of the site, copper and smaller quantities of platinum and palladium are also produced. The mine is operated by Toronto based Xstrat Nickel, part of the Swiss Xstrata group, with a proportion of profits being given to the Makivik Corporation of local Inuit communities. As with Red Dog, all workers have to be flown to the site. Production at the mine is about 1 million tons annually, but this is expected to rise to 1.5 million tons by 2011 and to 2 million tons by 2013. In part the increase in production will be from an extension of the mine south towards Lake Pingualuit. This 3.5km (2 mile) diameter, 250m (820ft) deep, almost circular lake formed in the crater of a meteor which struck Earth about 1.5 million years ago is claimed to hold the purest water in the world. There are plans to make the lake the centre piece of a Provincial Park. The mine extension could be allied to this proposal if mine transport could be upgraded to aid visitors to reach the lake, but there are fears that leaching from mine workings and waste might also contaminate the lake. Any extension of the site would require a renegotiation of the Impact and Benefits Agreement (IBA) with the local Inuit community, which was signed when the mine opened. The IBA included assistance with local education in an endeavour to ensure that Inuit constituted 20 per cent of the mine workforce. At present that figure has almost been achieved.

The Red Dog and Raglan mines in the Nearctic have been the subject of serious concerns for conservationists, but neither mine has produced the ecological damage for which a series of metal workings in Russia has been responsible. Visitors to the extreme north-east corner of Norway who take a night-time drive along the road south to the Øvre Pasvik National Park, which stands close to the point where Norway, Finland and Russia meet at a single point, will see a red glow in the eastern sky, caused not by the aurora borealis, but by Nikel, a Russian smelter complex close to the Norwegian border. Nickel deposits were discovered in the 1930s when the area was part of Finland. The USSR gained the site and the surrounding country in 1944 under the terms of the Moscow Armistice which ended the Soviet-Finnish war, and changed its name from Kolosjoki. Production of nickel at the site initially used local ore deposits, but after 1974 ore imported from Noril'sk, which has a much higher sulphur content, led to a huge increase in sulphur dioxide emissions, with significant ecological impact locally (and occasionally across the border in Norway, though the prevailing wind tends to take the emissions further into Russia). Total annual emissions are believed to have been in excess of 400,000 tons annually in 1979, though this figure decreased below 300,000 tons in the 1990s. The gas is considered responsible for the high levels of atopic disease in residents, particularly children. In a plume area downwind of the smelter, acid rain has caused the large scale destruction of vegetation, leaving a barren landscape, while vegetation outside this dead area shows high levels of heavy metals. A similar plume area has been observed downwind of the Zapolyarnij and Monchegorsk smelters, which are also situated on the western fringe of the Kola Peninsula. Monchegorsk is the largest of the Kola smelters.

Noril'sk is a much bigger smelting complex. Copper is known to have been worked in the area in the seventeenth century, but commercial mining/smelting did not start until 1935. At that time much of the work carried out at the site was by prison labour, Noril'sk being a prison camp, part of the Soviet Gulag system. The city close to the mine workings is the second largest in the Arctic (after Murmansk, which actually lies further south). Noril'sk lies at the southern fringe of the Taimyr Peninsula and has an extremely harsh climate for such a large and industrially important city, being snow covered for 250 days each year and having an average temperature of -15.8°C. Temperatures as low as -72°C have been recorded. The city also experiences Arctic winter (i.e. when the sun does not appear above the horizon) for six weeks. The city's wealth is founded on the local nickel, copper, palladium and platinum deposits, which are believed to be the largest in the world, with known reserves of at least 500 million tons. Ore extracted from five underground mines is processed in two ore concentrators and three smelters. Both concentrate and metal are transferred by rail to Dudinka, a port on the lower Yenisey River, which can be reached by sea-going vessels operating on the Northern Sea Route. On average Noril'sk produces 425,000 tons of copper, 240,000 tons of nickel, 90 tons of palladium and 20 tons of platinum annually, as well as significant quantities of cobalt. Currently the smelter produces 35 per cent of the world's palladium, 25 per cent of the platinum, 20 per cent of the nickel, 20 per cent of the rhodium, and 10 per cent of the cobalt. Ironically, platinum and palladium are used in the manufacture of catalytic convertors installed in motor vehicles to reduce pollution.

Heavy metal contamination in the Noril'sk area has frequently been cited in literature as a considerable problem, but sampling of ptarmigan at various points throughout the Arctic showed that the birds in Taimyr carried only 20 per cent of the lead burden of birds in Alaska and Canada, probably due to small scale gold mining and the use of leaded petrol. Similar trends were also seen in the mercury and cadmium burden of caribou/reindeer, with North American animals carrying levels almost 10 times higher than Eurasian animals. The worldwide ban on the use of lead in petrol will reduce metal burdens in all species. However, it has recently been discovered that the Arctic is a global sink for mercury, the suggested mechanism involving the polar sunrise which triggers the formation of elemental mercury which is deposited on snow. Atmospheric mercury derives from fossil fuel burning, cement production and waste disposal.

Extracted ore is smelted in Noril'sk, a process which generates massive pollution, chiefly in the emission of sulphur dioxide. The resulting acid rain downwind, together with pollution by heavy metals, has caused ecological damage on a massive scale: one claim which has been made is that the soil close to the smelter is so contaminated with palladium and platinum that it could be worked on a commercial basis. Beyond the zone of vegetation killed by acid rain, the tundra, on which reindeer forage, is heavily contaminated. The pollution also affects workers and others in the city, health problems similar to those of Nikel being prevalent.

The Russian operators of the site, Norilsk Nickel, whose value is estimated at in excess of £10,000 million, accepts that the pollution situation is unacceptable and has drawn up plans to reduce sulphur dioxide emissions by 70 per cent by 2020. At present the company estimates that almost 2 million tons of the gas are emitted annually. Despite the company's claim, environmentalists are dubious, pointing out that if sulphur dioxide emission is prevented, the quantity of sulphur which will build up at the site will create its own problems as the city is so remote that selling it will be almost impossible. Environmentalists also point out that there is a lack of independent scrutiny in Russia, while foreign scrutiny is not possible: Noril'sk is closed to foreigners, in part because it is apparently ringed by missile silos. However, the problem has to be seen in context. It is estimated that China is now emitting a total of 25 million tons of sulphur dioxide annually, and US emissions were about 30 million tons annually in the 1980s. There are also natural sources: the Kamchatka volcanoes emit 1,000,000 tons per year. The Noril'sk picture is not a happy one, but it is still the case that the majority of sulphur dioxide in the Arctic comes from non-Arctic industries.

It should also be noted that although the pollution plumes at Nikel and Noril'sk are horrendous, similar polluted plume areas would have occurred close to the early industrial sites of other areas of Europe such as northern England and the Ruhr, though there the pollution would have been less visible due to the absence of large areas of virgin forest, these having already been cleared. The difference is that the problems of industrial emissions is understood, leaving only the problem of reconciling profit with worker and environmental damage. In rapidly industrializing nations a reasonable balance is not always achieved.

Until the twentieth century diamond mining was carried out in places far removed from the Arctic or sub-Arctic, but recent finds have suggested the possibility that the area might be a rich source of the gems. In the 1950s diamonds were found in Yakutia, near Archangelsk and near the Yenisey River in Russia, making the country one of the world's leading suppliers of the gem. Currently there are several open pits operating in Yakutia. The pits, operated by the Alrosa company, produce 95 per cent of Russia's diamonds and account for about 25 per cent of world production.

In Canada two mines opened in the last years of the twentieth century. BHP Billiton's Ekati mine was the first to open, in 1998. It is in North West Territories, 300km (185 miles) to the north-east of Yellowknife (and 200km (125 miles) south of the Arctic Circle). For about 10 weeks each winter an ice road connects the mine to Yellowknife. By 2008 the mine – a series of four open cast pits and two underground mines – had produced 40 million carats (8 tons) of diamonds: it is expected to be active for at least twenty-five years. Ecologists have expressed concerns over the mine, as its footprint is relatively large (over 1,475 acres) and it lies on the migration route of the Bathurst caribou herd, one of North America's largest. Rio Tinto's Diavik mine lies about 25km (15½ miles) south-east of Ekati and began production in 2003. Peak production from both underground and open cast workings is expected to be 8–10 million carats annually with an expected total production of 110 million carats over a life of about twenty years. As with Ekati, the mine is connected to Yellowknife by a winter ice road, but problems with late freezing and early thaws (an indication of the effect of global warming) has meant Rio Tinto's interest in constructing a 200km road to Bathurst Inlet and construction of a port there, something which environmentalists oppose because of the probably negative effect on the Bathurst caribou. The mines projected full operation footprint is expected to be 2,470 acres. The waste from each mine will be considerable as extraction rates are not expected to exceed 3 carats per ton, i.e. to extract 110 million carats the Diavik mine will need to process almost 4 million tons of rock. The extraction process, which involves crushing the diamond-bearing rock, is largely gravity-based, but does involve significant quantities of water.

The Smoking Cliffs in Franklin Bay, on Canada's northern mainland east of the Mackenzie Delta, were named by John Richardson during John Franklin's second land expedition in 1825–7. The cliffs are composed of bituminous shale which, it is believed, was ignited by a lightning strike several thousand years ago and have been smouldering ever since. The Inuit must have known the area, and it may even have given them the idea of using oil which seeped to the surface at several places in the Arctic as a fuel as it is known that they compacted oil-stained soil into bricks and used it for cooking and heating. The Dene Indians who lived to the west of the Great Bear Lake also utilized oil seepages, spreading it over their skin canoes to improve the waterproofing. The use was noted by Alexander Mackenzie when he travelled the river, but not until 1914 was any attempt made to develop the oil commercially. With commercial development came the building of a town, Norman Wells. During the 1939–45 war the United States constructed a pipeline (the Canol line) to take the Norman Wells oil to Whitehorse where it would meet the recently completed Alaska Highway, but the poor understanding of such pipelines meant it was never very successful: today about 25,000 barrels of oil are pumped south daily along a pipeline to Zama City in Alberta.

The most famous of the Arctic oilfields is that at Prudhoe Bay on Alaska's North Slope where oil was discovered in 1968 (though oil did not flow from the field until 1977), though the history of oil exploration on the North Slope is much longer. The first exploration was in 1922, and the following year the Naval (the National since 1976) Petroleum Reserve was designated. Covering 23 million acres, the reserve is the largest area of unprotected public land in the United States. The reserve is estimated to hold between 5,000 and 12,000 million barrels of oil. Drilling licences for the Reserve have been issued, but not without significant objections from conservation groups.

When it was decided to open the Prudhoe Bay oilfield, the difficulties of shipping oil from the field led to the building of the 1,300km (800 mile) pipeline from Prudhoe to Valdez, an ice-free port on Alaska's southern coast. Construction of the pipeline was a considerable feat of engineering as the oil is pumped at high temperature and there was a need to ensure that heat losses did not melt the permafrost. (Ironically considering the effort which went into the design and construction to avoid melting the permafrost, the permafrost is now melting, sinking ground causing problems for the line.) The pipe runs mainly above ground which gave conservationists concerns as it traverses a caribou migration route. Generally the pipe is raised sufficiently high that the animals can walk beneath it, though some older sections are too low. However, many caribou show a marked reluctance to go anywhere near the pipeline or other oilfield related structures and there is some evidence that the interruption of the normal migration routes has resulted in a population decline. At Prudhoe Bay, workers are accommodated in the village of Deadhorse, where the construction of administration offices, buildings and roads caused changes to local hydrology which led to flooding, though, in the main, environmental damage in the immediate area has been minimal. Studies have shown that brown bears and Arctic foxes whose territories include rubbish dumps associated with the oilfields raise more and healthier cubs/pups than those whose territories do not (strangely this has nothing to do with an enhanced diet, but with the fact that female animals do not have to travel so far to find food, travel occasionally leading to the deaths of young from drowning etc or predation). Not surprisingly, prey species of the bears and foxes (birds and rodents) show population decreases close to the dumps.

At discovery the Prudhoe Bay field was considered to contain 25,000 million barrels of oil (though not all of this is recoverable with the present status of technology). Average daily throughput of the pipeline is about 1 million barrels (but has been as high as 2 million barrels). At any one time the pipeline holds about 9 million barrels: travelling at 6–7km per hour, the oil takes eight or nine days to reach Valdez.

The Prudhoe Bay fields also hold a considerable quantity of natural gas (at least 35 trillion cubic feet (1,000 cubic kilometres), perhaps three times that quantity if all likely on-shore and off-shore reservoirs are included). At present some of this is extracted to run gas turbines that power the Prudhoe complex: the remainder is pumped back into the underground reservoir. It is planned to build a second pipeline to transport gas to Valdez, from where it can be shipped as a liquid (though there are competing projects which suggest a pipeline across the border into Canada to link with the gas pipeline there, or one heading south and then following the Alaska Highway). Although the gas reserve seems enormous, it is worth remembering that the annual demand in the US is 25 trillion cubic feet.

Two years after oil began to flow from the Prudhoe Bay field the first leases were granted for off-shore drilling in the Beaufort Sea off Alaska's coast. These were followed in 1989 (the year of the *Exxon Valdez* disaster) by the first exploratory drilling in the Chukchi Sea. Oil first flowed from a Chukchi Sea well (Northstar) in 2001. In early 2008 the Chukchi Sea became the centre of a debate when the Minerals Management Service (MMS), a section of the US department of the Interior, decided to issue leases for drilling (Lease Sale 193) which covered over 30 million acres. Just a few days after the decision the US Fish and Wildlife Service, a different section of the same Department, announced that it would not be able to meet the legally required deadline for a decision on whether to place the polar bear on the list of threatened or endangered species under the Endangered Species Act. In part the delay was caused by complaints from the State of Alaska, which submitted evidence that the population of polar bears was 'abundant, stable and was not threatened by direct human activity' and that to allow threatened status would 'deter activities such as ... oil and gas exploration and development'. The underlying scientific evidence was roundly criticized by other scientists as based on out-of-date and flawed data. The State Governor, Sarah Palin, later John McCain's chosen Vice Presidential candidate in his failed bid to become US President in 2008, has made a number of statements doubting global warming or, at least, doubting a human dimension to it, and at one time claimed that the building of a gas pipeline across Alaska was 'the will of God'.

Not surprisingly, both Alaskan native and conservation groups saw the decisions as linked, accusing the administration of President George W. Bush (aided by the State of Alaska) of deliberately stalling the polar bear decision so that the leases could be issued. The groups point out that the Chukchi Sea is not only an important habitat for the bears, but for seals, walruses, bowhead whales and migrating birds, and that MMS's own study of the likely environmental impact of oil production in the sea suggests a 40 per cent chance of a major spill during the life of the field. Such a spill, the groups claim, in combination with the reduction in sea ice cover as a consequence of global warming, would represent a serious threat to the continued survival of polar bears and other Arctic species. Protests that the awarding of leases was in contravention of the 1976 International Agreement on the Conservation of Polar Bears and their Habitat (Appendix 3) were not heeded. Although Article II of the Agreement discusses feeding sites and migration patterns, there are no specific references to the effects of climate change, pollution, industrial development and tourism. Although the casual reader might consider these issues to be covered by the spirit of the Agreement, lawyers can use the lack of specifics to defend all manner of activities.

In Russia the presently known oil and gas reserves are in western Siberia, in an area bounded by the White Sea to the west and the Taimyr Peninsula to the east, and extending north under the Barents Sea. At present the richest fields are in the Timan-Pechora Basin in the Nenets Autonomous Okrug. (NAO – the *okrug* (district) covers the northern slice of the Russian mainland from the eastern edge of the White Sea to Baydaratskaya Guba on the Yamal Peninsula's western edge.) Development there began in the 1930s, but was then halted for forty years before resuming. Oil and gas were discovered in the Yamal-Nenets Autonomous Okrug (YNAO – which includes the Yamal Peninsula, the eastern shore of the Obskaya Guba, and the mainland south of the bay) in the 1960s, the decade also seeing confirmation of both oil and gas in the NAO. Gas production began in the YNAO in the early 1970s, and both oil and gas production in the NAO ten years later. Workers were sent from across the USSR to work on the pipeline to carry the gas, and the construction cost a vast amount of money. To recoup the money the government sold the gas to Western Europe, but announced that it was actually being exported to Poland.

The break up of the Soviet Union reduced production, the newly emerged Russian Government needing time to collect its thoughts. It soon become aware of the economic importance of exports (particularly of gas) and production of both oil and gas increased. At that time, the pipeline was privatized, sleight of hand allowing it to become the property of former communist officials, with the former Minister of the gas industry being appointed as CEO. But gas was merely one example of the way in which privatization was carried out in the new Russia. Harking back to former times, ordinary Russians, who had been brought up under communism to understand that they owned the gas finished up with nothing, As one example, Russians currently pay more for oil and gas than Americans do, despite the vast Russian reserves of both and the disparity in income levels between the two nations.

Russia has the largest proven gas reserves in the world and is also the world's biggest exporter of gas. Gazprom, Russia's largest company, operates a pipeline over 150,000km (93,000 miles) long, which feeds gas to neighbouring states and Europe. Further development of the Timan-Pechora Basin by a multinational group of companies was announced in 2004, but the more recent announcement by Gazprom that it was intending to develop the Shtokmanovskoye field under the Barents Sea on its own was greeted with shock by US and Norwegian companies, particularly as Gazprom also stated that it was reconsidering its decision to sell liquefied gas to the US – shipped in tankers from Murmansk, Russia's only Barents Sea port which is ice-free throughout the year. Shtokmanovskoye is believed to be the world's largest gas field, with reserves of over 140 trillion cubic feet (almost 4,000 cubic kilometres). However, as the venture is the first open-sea drilling project on such a scale, Gazprom has now agreed a joint venture with Total (France) and StatHydroOil (Norway). The first platform began construction in Vyborg in July 2008.

Other important Russian oil and gas fields include Yuzhno-Russkoye in the YNAO with estimated gas reserves in excess of 40 trillion cubic feet (about 1,000 cubic kilometres). Gazprom is also developing the 'Mega Yamal' project at the Bovanenkovo gas field, with a new pipeline to be built from the Yamal to Ukhta, a town in north-west Russia. During the summer of 2008 the gas pipeline crossed Baydaratskay Bay; the line is expected to be completed in 2011.

Although there have been problems between the native reindeer herders and the oil and gas companies, in general relationships have been good, with the companies agreeing to buy reindeer meat from the herders in preference to ferrying meat to the drill sites. The company rents the herders a slaughterhouse it has constructed, and bringing meat in from outside is expensive, while the herders have a guaranteed outlet without the necessity of driving their herds to market, so the arrangement suits both sides. The Gazprom workforce has become the only market for the Nenet's reindeer meat. That this is the case can be gauged from the YNAO 'coat of arms' including polar bears, reindeer and a drilling rig complete with gas flare. But while relationships with the locals have been reasonable, ownership and development of the field has been the subject of dispute after a subsidiary of Gazprom signed a contract with the US company Moncrief Oil International, which Gazprom refused to recognize. Gazprom regained control of the field, but is developing it in collaboration with German companies. The Priobskoye oilfield was discovered in 1982 in the Khanty-Mansiyskiy Autonomous Okrug, south of the Yamal-Nenets region. The field embraces both banks of the Ob River and extends over 5,000km^2 (12 million acres). There are also other important fields located in sub-Arctic Russia, including the Samotlor field near Lake Samotlor in the Urals where a decision in the 1960s to use western technology and Texan workers, injecting water into the field in an attempt to maintain production as yield was falling, led to mixing of water and oil, and reduced recovery efficiency. Newer western technologies are now allowing extraction rates to be improved. Other important large fields are Lantorskoye and Oskoye. There have also been attempts to find oil in the Anadyr River delta, with unknown results. There are rumours that oil was found there, but the economic difficulties of extraction have postponed exploitation.

In Canada the first exploration of what was to become a major gas field, in the Mackenzie Delta, was made in the late 1960s. However, exploration soon ceased as difficulties with native groups and conservationists raised doubts whether a pipeline to export the gas would ever be constructed. With liquefaction and export by tanker having severe technical difficulties, interest moved elsewhere, gas being produced first from a site in the Painted Mountains close to the Nahanni National Park in the south-western corner of the North West Territories (NWT). The field has now produced over 3 trillion cubic feet of gas. There was renewed exploration in the Mackenzie Delta in the 1980s, but gas was first exploited commercially only in 1999 when it was pumped to communities along the northern Mackenzie River. The Mackenzie Gas Project proposes to expand usage into southern Canada by constructing a 1,220km (758-mile) pipeline along the Mackenzie Valley to link with an existing gas pipeline in north-west Alberta. The proposed route of the pipeline crosses four native people areas. Negotiations with these groups, together with objections tabled by conservation groups mean that construction of the pipeline is unlikely to start before 2010.

In 1974 the Bent Horn oilfield was discovered on Cameron Island, one of the smaller Parry Islands, to the north-west of Bathurst Island, in what was then NWT, but is now part of Nunavut. In 1985 the first oil was shipped by ice-breaking tanker to a refinery in Montreal. Shipments continued until 1996 by which time about 12 million barrels of oil had been extracted.

Oil and gas reserves have also been identified and developed in the Norwegian Sea. The Ormen Lange field is the largest gas field in Norwegian waters, the gas being piped along a 1,200km (745-mile) seabed pipeline to Britain. The field is estimated to have reserves of over 10 trillion cubic feet. In Arctic waters, in 1984 the Snøhvit field was discovered in the Barents Sea about 140km (87 miles) north-west of Hammerfest. Though primarily a gas field, with estimated reserves in excess of 6 trillion cubic feet, the field also holds over 100 million barrels of oil. The plan to develop the field was highly controversial in Norway, a country renowned for its love of wilderness areas and its environmental conscience. Conservationists argued that the Barents Sea was too vulnerable an area for industrial development and that the on-shore processing plant – it was planned to pump the gas ashore to be liquefied for overseas shipment – would significantly increase Norway's carbon dioxide emissions (by almost 1 million tons annually, an increase of about 2 per cent) at a time when the country should have been setting an international example by reducing them. Angry protests saw attempts at preventing construction work at the processing plant site, with the police arresting protesters. Despite the protests, and the national debate the protesters aroused, gas production from the field began in 2007, and the gas was piped ashore. In April 2008 the field operators (Norway's Statoil) announced that carbon capture had begun at the site, carbon dioxide extracted from the gas at source being pumped back into the field beneath the gas-bearing rock. Statoil claim that 700,000 tons of carbon dioxide will be captured annually.

In 2000 the Goliat oilfield was discovered, with estimated reserves of 250 million barrels. The decision not to pipe the oil to the mainland, but to process and ship it directly from the platform (or from the nearby small, uninhabited island of Soerøya) led to protests in Hammerfest where the jobs (and money) a processing plant would have created were anticipated. The field operators justified their decision on the basis of a very limited market for oil in northern Norway and the expense of building the underwater pipeline. Production from the Goliat field is likely to make the extraction of oil from the Snøhvit field economically attractive. Following the successful discoveries in the Barents Sea, despite further protests, Norway issued licences for new drilling in 2006 and expects to see the development of further fields.

In the 1970s the first licences were issued for gas and oil exploration in the waters off Greenland's western coast in an area south of Disko Bay. One relatively minor discovery was made: an on-shore drilling programme begun in 1996 has been equally unsuccessful so far. As a consequence, little further interest has been shown by exploration companies despite the best efforts of the Greenlandic parliament to make licence acquisition attractive.

In the 1970s seismic data was collected in the waters off the Faroe Islands, encouraging results leading to exploratory drilling in 1982. Further seismic studies were then made, and further drillings were made in 2002, these encouraging the Faroese Government to issue exploration licences in 2005. The outcome of the Faroese studies have been closely observed by the Icelanders: the collapse of the Icelandic banks in the international banking crisis of 2008 has, of course, produced a financial imperative in Iceland and seismic studies have now begun.

By 2008 Russia's Arctic oilfields had produced around 80 billion barrels of oil and 500 trillion cubic feet of gas (in energy terms, a barrel of oil is equivalent to about 35,000 cubic feet of gas). Canada's fields had produced 250 million barrels and one trillion cubic feet, while 16 billion barrels of oil had been extracted from Alaska's North Slope, though production there is now falling. No production figures are yet available for Norway's Barents Sea fields, but the Norwegian Sea fields have so far produced about 3 billion barrels of oil and some 3 trillion cubic feet of gas.

A male polar bear on the sea ice between Spitsbergen and Edgeøya. He is a little over two years old and has just parted company with his mother.

Chapter Six
The Future

A Yupik hunter on St Lawrence Island once told me that what traditional Eskimos fear most about us is the extent of our power to alter the land, the scale of that power ... Eskimos, who sometimes see themselves as still not quite separate from the animal world, regard us as a kind of people whose separation may have become too complete. They call us, with a mixture of incredulity and apprehension, 'the people who change nature'.
Barry Lopez, *Arctic Dreams*

As the twentieth century drew to a close, fears began to be raised about the long-term security of world energy supplies. As the economies of China, India, Brazil and other developing nations expanded, their energy requirements increased. The developed world looked on in alarm as energy supplies became more expensive and both the supply of fossil fuels from known sources and the identification of new sources showed signs of slowing down. Inevitably the eyes of the developed world turned towards their northern boundaries. But what Arctic was revealed by those covetous glances?

The Arctic is vulnerable to pollution, and warm air and ocean currents flow from the tropics, adding their own pollutants to those from local metal mining and smelting, and drilling for fossil fuels. Arctic rivers flow from temperate areas to the northern ocean, bringing agricultural chemicals, pesticides and industrial pollution. All pollutants make their way north, but POPs (persistent organic pollutants) present a particular problem for Arctic lifeforms. POPs do not easily dissolve in water, but dissolve readily in lipids – fats and oils such as blubber and mammalian milk and therefore become part of the biological chain. The adaptations of Arctic wildlife and its peoples to their environment therefore make them especially vulnerable to these pollutants. This is particularly true for animals that rely on fat reserves to survive the winter. The Arctic's climate also makes it more susceptible to POP pollution. The lack of photodegradation of the chemicals during the long Arctic winter allows survival rates higher than in southerly regions, while low temperatures inhibit the natural biodegradation of the compounds. Pollutant particles also attach readily to snowflakes, which have a large surface area relative to rain drops, and so are easily brought to the ground or ocean surface. Studies of one form of POP, PCBs (polychlorinated biphenyls) have shown that because they dissolve readily in fats and oils they are concentrated in the blubber of seals and other sea mammals, and consequently are present in very high concentrations in polar bears. Scientists call this bioaccumulation: the higher the position of an animal in the trophic pyramid, the higher the concentration of pollutants it will acquire in its lifetime. In relative terms, assuming the concentration in sea water is 1, the concentration in Arctic zooplankton in the worst affected areas would be 12,000, in fish 200,000–500,000, in seal blubber 500,000–1,500,000, and in polar bears, beluga and narwhal, up to 30,000,000. Because traditional food still forms a significant fraction of their diet, Inuit also show elevated PCB levels, this being particularly worrying in nursing mothers. The startling conclusion of one study was that some of the tested Inuit had such high concentrations of PCBs that their bodies would be classified as hazardous waste if they had to be disposed of as 'non-human' material. Other pollutants are similarly concentrated in Arctic native peoples and wildlife; one study indicated that the mercury levels in Greenlanders were the highest ever measured in humans.

Although the ozone hole in Antarctica receives more publicity, there is also a hole above the Arctic, with a correspondingly enhanced level of ultra-violet radiation to Arctic dwellers and wildlife. After the discovery of ozone thinning above Antarctica, research showed that the loss was primarily due to the presence of chlorine in the upper atmosphere. The source of the chlorine was identified as chlorofluorocarbons (CFCs), used as coolants in refrigerators (freons), in aerosol spray cans and other industrial products. The ozone holes are seasonal, forming when the poles emerge from the long polar night, and reaching a maximum during the polar summer. In 1985, in response to growing concern over the loss of the protective layer of ozone in the atmosphere twenty nations signed the Vienna Convention for the Protection of the Ozone Layer and accepted the need to control anthropogenic activities that caused the release of ozone-depleting chemicals. In 1987 these measures were incorporated into the Montreal Protocol on Substances that Deplete the Ozone Layer. Signatories of the Protocol now number more than 150. As a consequence of the Protocol the coolants in refrigerators and car and domestic cooling devices became more environmentally friendly. It is now believed that full recovery of the ozone layer at both poles will be complete by about 2030.

Of far more significance was the realization in the late-twentieth century that man's activities were not only increasing the levels of atmospheric pollutants and decreasing the levels of ozone in the polar regions, but was altering the Earth's climate. It is now accepted by all except a few die-hard deniers (chiefly politicians) that the average temperature of the Earth is increasing, and that the increase is related to the amount of carbon dioxide in the atmosphere which, in turn, is due to the burning of fossil fuels (as well as other processes allied to industrial activity). In making such a bold claim, it is, of course, necessary to qualify the statement by noting that, as with all other scientific theories, 'proof' of the claim that anthropogenic fossil fuel burning is responsible for global warming is not, strictly, possible. The theory could be disproved, but can only be set up as a working hypothesis which fits available facts, despite newspaper claims to the contrary.

In the absence of the atmosphere the Earth would be in thermal balance, the incident radiation from the sun being equal to that reflected back into space. Solar radiation arrives as short-wave radiation to which the atmosphere is, essentially, transparent. At the Earth's surface it is reflected as long wave radiation, which is absorbed by some constituents of the atmosphere – water vapour, carbon dioxide, nitrous oxide, methane and others – re-emitted and re-absorbed, resulting in a warming of the atmosphere. The Earth is still in thermal balance, but at a different temperature. With no atmosphere the average temperature on Earth would be -20°C: it is actually +15°C. Those constituents of the atmosphere responsible for this warming act in much the same way as the glass in a greenhouse, explaining why they are occasionally called greenhouse gases. The potential impact of carbon dioxide in the Earth's atmosphere has been recognized for over 100 years, but the fact that the gas' concentration was increasing has only been known since ice cores were taken in Antarctica and Greenland, allowing the concentration level to be measured, year on year, over a prolonged period.

In the Arctic, the level of warming is higher (almost double) the Earth's average because of a number of positive feedback mechanisms. The reflection of solar energy is high from ice and snow. If sea ice melts, the ocean that is revealed is dark and reflects much less radiation. If snow cover melts, the ground beneath is also much darker and so also reflects much less (and absorbs more energy). The reduction in sea ice cover and annual snow cover therefore means that the temperature of the atmosphere above the Arctic increases faster than in temperate zones where the change in reflectivity is much less stark. A warming Arctic means that the permafrost which underlies much of the area's landmass also warms. Permafrost stores both methane and carbon dioxide, so thawing releases more greenhouse gases. Warming of the Arctic has also meant an increase in precipitation in the area, with an enhanced fall as rain rather than snow. Nothing clears snow cover, or melts sea ice, as efficiently as rain, leading to another positive feedback mechanism to enhance Arctic warming.

As the Earth warms the climate is becoming unstable, with an increase in 'freak' weather events. The twenty-first century has also seen a succession of record reductions in the summer sea ice cover of the Arctic Ocean. The winter temperature of the Arctic is now 3–4°C warmer than it was in the mid-twentieth century. The sea ice is not only decreasing in area, but in thickness: it is now, on average, 30 per cent thinner than it was ten years ago. Though the area of winter sea ice in 2008–9 was larger, the volume was smaller: the six lowest volumes of winter sea ice ever recorded have occurred in the last six years. Loss of the sea ice would be disastrous for the polar bear and many Arctic sea mammals, but it has led to easier travel in the area. And easier travel means easier exploration and will mean easier exploitation of the area's mineral resources.

In 2008 it was announced that the CO_2 level in the Earth's atmosphere was at its highest level for 650,000 years. More worryingly, despite all the concerns expressed, the meetings held and the agreements reached, the mean growth in CO_2 level had increased from an average of 1.5ppm (parts per million) per annum in the period 1970–2000, to 2.1ppm per annum in 2000–7. The increase in 2008 was anticipated to be 2.14ppm. In July 2008 the USA finally signed up to a G8 proposal to cut CO_2 emissions by 50 per cent by 2050. The move was seen as very positive, but within hours of the announcement, Brazil, China, India, Mexico and South Africa demanded an even bigger reduction to offset the growth of their own industries. Then, in December 2008 the twenty-seven countries of the European Union (EU) agreed an overall cut of 20 per cent of 1990 CO_2 emissions by 2020, with each country having its own limit and an EU-wide carbon trading scheme intended to reduce the burden of cuts on heavily polluting industries, something which was insisted upon by the developing nations and by Germany, which produces much of its electricity by burning coal. The politicians were very pleased with the outcome, French President Nicholas Sarkozy, whose six-month term as head of the EU was in its final weeks and who had staked his reputation on getting a deal, claimed the meeting, at Poznan in Poland, would go down in the history of Europe. Environmentalists were much more muted in their praise, considering that in effectively excluding the big polluters, the deal had done little to ensure a real reduction in emissions.

The Arctic Basin

Antarctica is a landmass surrounded by sea and differs fundamentally from the Arctic, which is sea surrounded by land. On that land the search for minerals has begun. On Svalbard the Store Norske group (the Norwegian coal mining company) has used the provisions of Svalbard Treaty and Mining Code to make over 300 claims, most in the central area of Spitsbergen. Claims are controlled by the Svalbard Mining Inspector, and for an annual fee of 6,000 Norwegian kroner (about £625 or $875 at early 2009 rates) any company of a Treaty signatory country can claim an area of about 10km^2 and then has the sole right to mine within that area. Though the majority of the Store Norsk claims are in areas where coal is likely to be found, the company is aware of the potential for copper, gold, lead and zinc, as well as oil. It is likely that in the future Russian companies will be equally likely to take an interest in Svalbard's potential for minerals. On Greenland, those leading the debate for independence from Denmark hold out the possibility that Greenlanders might grow rich on diamonds, gold, lead and zinc, and mining companies are already prospecting easily accessible areas. In Canada, Russia and the US, mining companies are also sending prospectors northwards in search of bauxite, copper, diamonds, gold, manganese, molybdenum and nickel. But in those nations the main area of interest is not metal or gems, it is oil and gas, and the likelihood of finding significant quantities in the ocean basin area that forms the centre of the Arctic.

The seabed beneath the Arctic Ocean is not uniform, comprising a series of basins 3,000–4,000m (1¾–2½ miles) deep separated by ridges. The Mid-Atlantic Ridge, the existence of which has created Iceland and Jan Mayen, separates Greenland and Svalbard, forming the 1,500km (930-mile) Nansen-Gakkel Ridge, which rises 2,000m (5,400ft) above the seabed. The seismically active area of Siberia's Verkhoyanskiy Mountains is on a continuation of this ridge (though the mountains also lie between tectonic areas of the Siberian platform and the micro-continents of far eastern Siberia). Running parallel to the Nansen-Gakkel Ridge are the Lomonosov Ridge, which lies beneath the Pole itself, and the Alpha Cordillera/Mendeleyev Ridge, which forms an arc extending from northern Ellesmere Island to Wrangel Island. The Lomonosov Ridge is even larger than the Nansen-Gakkel, extending for 1,800km/1,100 miles (between the North American continental shelf near Ellesmere Island/Greenland to the Eurasian continental shelf near the New Siberian Islands) and rising 2,000m/9,840ft above the seabed. Between these extensive ridges are the deep basins: the Nansen Basin, between Svalbard/Franz Josef Land and the Nansen-Gakkel Ridge; the Fram (or Amundsen) Basin between the Nansen-Gakkel and Lomonosov ridges; the Makarov Basin between the Lomonosov and Alpha/Mendeleyev ridges; and the Canadian Basin between the Alpha Cordillera and Alaska/Canada. Sampling of the basins suggests they are from 60–135 million years old and that they were indeed formed by sea floor spreading. The Fram Basin is the deepest, at almost 4,500m/14,750ft. The deep basins are ringed by a series of much shallower seas, which lie above the continental shelves. The Arctic sea floor is the slowest spreading floor on Earth, spreading four times slower than the Pacific, ridge volcanism being limited (though extremely variable, as recent studies of the Gakkel Fidge have shown).

The continental shelves of Eurasia are extensive, reaching to and beyond Svalbard and the islands of Arctic Russia. The shelves occupy 35 per cent of the area of the Arctic Ocean, yet account for only 2 per cent of the water volume. The Siberian shelf, which forms part of the Eurasian shelf, is the world's widest, being up to 900km (560 miles) wide. The seas above the Eurasian shelf are very shallow, never exceeding 100m (330ft), and often being only 10–30m (30–100ft) deep. The seas are also warmer than might be expected. The Barents is warmed by the Gulf Stream, and while the stream's influence wanes sharply to the east, the eastern seas are warmed by the seasonal discharge of warm fresh water from Russia's huge rivers. To the west of the Barents Sea the deeper Greenland Sea (more than 2,000m/6,500ft deep) separates Greenland and Svalbard. The sea represents the widest breach in the continental landmasses surrounding the Arctic Basin.

The North American continental shelf is generally less extensive than that of Asia (though as the shelf underlies the Canadian Arctic islands that suggestion can be questioned). The seas of the American Arctic are therefore deeper: Baffin Bay, between Greenland and Baffin Island, is over 1,000m/3,250ft deep, as is the Beaufort Sea (which is over 3,000m/9,850ft deep at its northern extreme). It is interesting to note that the continental shelf beneath the Beaufort Sea is much less extensive than that beneath the Chukchi Sea which borders it.

The sea ice cover of the basin – both the permanent, summer, ice and the winter ice – are subjected to a number of currents, which cause constant drifts to be added to the temporary movements of local winds. The Transpolar Drift moves ice from the East Siberian Sea towards the Denmark Strait, while the Beaufort Gyre is a clockwise current in the Beaufort Sea. In addition to these large-scale currents, there are smaller currents and gyres associated with the coastlines of North America and Eurasia.

While still in his position as President of the US arm of Royal Dutch Shell, John Hofmeister (who retired at the end of May 2008) stated that Shell believed 'that the most prolific remaining conventional oil and gas resources are in the Arctic or sub-Arctic, because we've pretty much developed the geologies south, whether that's in the United States or whether that's in Europe or Asia.' The rest of the oil industry agrees with him. The logic is flawless: if you have searched everywhere else, then the most likely place you will find something is the place you haven't looked. Some studies indicate that the Arctic Basin is indeed a likely source of oil, the discovery of fossil azolla fern species in drilling sites on opposite sides of the basin leading to the assumption that at one time in its history the basin was a landlocked sea in which organic sediments from the ferns, and other plant life, accumulated. Figures of 400 billion barrels of oil are often seen, consistent with the estimated reserves beneath the Chukchi Sea which the US MMS states are 15 billion barrels of oil, together with over 70 trillion cubic feet of gas. But there are dissenters, those who believe that the Arctic Basin is unlikely to be that prolific, and that there is a large uncertainty in the much more easily defended figures for the Chukchi reserves. Away from the Arctic Basin the main areas of interest in the Arctic are central and east-central Siberia, the Barents Sea, the Labrador Sea, Canada's Arctic islands and Hudson Bay and the eastern Bering Sea. The waters off Greenland still attract some attention, but the hope seems to derive rather more from the gross similarities of the area with the North Sea. Claims for oil in Icelandic waters are similarly based, though this time on the hopes of oil companies for success near the Faroes.

When considering the likelihood of the Arctic holding prodigious quantities of oil, the question naturally arises whether it should be extracted. John Hofmeister, the US Shell boss already quoted above, believes it is essential that the US exploits its Arctic reserves so as to reduce its reliance on foreign energy supplies. Yet Hofmesiter believes that global warming is a man-made problem caused by an increasing carbon dioxide concentration in the atmosphere and is in favour of a cap on US carbon dioxide emissions. The dichotomy this position exposes is reflected elsewhere, not only among the executives of oil companies, but at the higher echelons of government. In many western nations lip-service is paid to the needs of greenhouse gas reductions, while policies are pursued which favour economic growth. Politicians do not want to lose their jobs any more than the rest of us do and know, with certainty, that calling for a decrease in living standards will lead to electoral defeat. For company bosses, so long in thrall to shareholder desires and boardroom requirements, the decisions are no easier.

But let us, for the moment, ignore the arguments over the desirability of extracting oil and gas (and other minerals) from the Arctic, ignore the consequences for the Earth as a whole and consider only the implications for the area.

Sea ice represents a problem for off-shore gas or oil platforms. Moving sea ice can damage the platforms themselves: anyone who has travelled on sea ice will know how big pressure ridges, formed by the action of currents on the ice, can be, an indication of the power of ocean currents and the unlikely malleability of the ice itself. As well as throwing ridges of ice upwards, the pressure can throw them downwards – as surface ridges are occasionally called 'hummocks', these sub-surface ridges have been christened 'bummocks'. Bummocks may gouge trenches in the seabed, damaging pipes laid on the bed or in shallow trenches. Avoiding bummocks means cutting pipe trenches very deep, which is expensive. The action of freezing and thawing can also distort the seabed which also damages pipelines unless they are at a considerable depth. Even more deadly to both platforms and pipelines would be an ivu. Ivu is an Inuit word describing the sudden movement, at speed, of a mass of sea ice, which can actually travel considerable distances on to shore. (The phenomenon is occasionally called an ice tsunami, or an ice shove.) Scientists were sceptical about the existence of the phenomenon but there is now overwhelming evidence from across the Arctic for its existence: for example in 1982, when the bodies of a family of five apparently overwhelmed by an ivu were discovered near Barrow, Alaska. Ivus are now accepted as real: the production method is still unclear, though a theory based on coastal flooding caused by on-shore wind and current has been proposed. An ivu would be a considerable danger to an oil or gas platform, and very difficult to protect against. A further hazard would be the existence of unidentified seabed pingos. Pingos are conical masses of ice formed in permafrost areas which can grow to considerable heights (as high as 50m/160ft). Since pingos are a permafrost phenomenon they can also exist beneath the sea. A large, undetected pingo would be a hazard to shipping. In addition underwater pingos are believed to act as a conduit for highly flammable methane, adding to the hazard.

Oil and gas in the Arctic is expensive to extract. Sea ice presents particular technical difficulties, but so too do extreme seasonal climatic changes. Usually there is also a very limited, perhaps non-existent, local labour, adding to the transport costs and salary demands of workers who must leave families behind. And the extracted fuel must then be transported over long distances as there will be an extremely limited local market.

Oil production in the Arctic also brings unique environmental problems. Although improving technology has reduced the footprint of the well heads (though the drive to do so was as much to do with economics as environmental concerns), the construction that inevitably accompanies oilfield development does considerable, if localized, damage. Wheeled and caterpillar-tracked vehicles create ruts that deepen and widen; construction causes changes in local hydrology (the construction of a 7km/4½-mile section of road near Prudhoe Bay caused the subsequent flooding of more than 320 acres of surrounding land), while oil production, including gas flares, releases hydrocarbons and other pollutants, including heavy metals into the environment. Land-based development can interrupt the migration routes of species (platforms at sea can do much the same). Developments can also fragment the range of species, something which can lead, ultimately, to a declining population as separated individuals cannot reach each other to breed. This has happened to tigers in India, where reserves are well separated, while the habitat in the corridors that once ran between them has been destroyed. It might seem unlikely that this could happen in as vast an area as the Arctic, but as the sea ice shrinks, so does the total range of a species such as the polar bear. Although many companies start with good intentions regarding the restoration of industrial sites, in the Arctic the regeneration time for disturbed landscapes is measured in decades. On Svalbard, it is still possible to see the ruts made by wheeled vehicles on the summer tundra before their use was banned, the rut edges still sharply defined many years after they had been cut.

From the point of view of wildlife, an operating and recovering landscape are similar – they represent lost habitat. Although, as noted earlier, some individual animals may gain from the existence of a site (or, to be specific, the site dump), in general the effect of industrialization is never positive and rarely neutral. Studies conducted by the Alaska Department of Fish and Game show that the female caribou productivity around the North Slope oilfields, including Prudhoe Bay, has declined since oil production began because of the interruption of their calving grounds. The caribou demonstrate a 3–4km (1¾–2½-mile) avoidance of the structures on the oilfields, including roads and pipelines. The likely effect on wildlife makes the decision by the Bush administration to view favourably drilling in the Arctic National Wildlife Refuge unfortunate. The Refuge comprises almost 20 million acres of north-east Alaska between the Brooks Range and the Beaufort Sea coast. When it was created (as the largest protected area in the United States) by the Alaska National Interest Lands Conservation Act in 1980, Section 1002 of the Act deferred a decision over 1.5 million acres of the coastal plain. It is this area, Area 1002, which is contentious. The debate over drilling in the Refuge has continued since its creation, though became highly topical in 2008 when George W. Bush not only backed drilling in the Refuge, but off-shore drilling as well, something his father had vetoed when he was President. Much of the argument is over the likely oil reserves in Area 1002, US Geological Survey estimates vary from as little as 5 billion barrels to as many as 16 billion depending on a number of factors and imponderables. Critics of the scheme point out that the projected daily maximum production of about 700,000 barrels over field lifetime of some thirty years has to be seen in the context of current US daily usage of over 20 million barrels. Proponents of exploitation note that only 2,000 acres of Area 1002 would be involved. However, that figure does not include land disturbed for pipelines, etc. or land not directly covered by site infrastructures, even if the land is wholly enclosed by pipelines, roads, etc. Critics note that at Prudhoe Bay the area equivalent to the 2,000 acres is 12,000 acres while the best estimate of the actual footprint of the oil production complex is over 600,000 acres. Native peoples close to Area 1002 (including some across the border in Canada) are also concerned that the proposed production site is on the migration route of the Porcupine caribou herd.

Of much greater importance to the environment is the possibility of an oil spill. Before the exploitation of oil and gas, oil seeped into the environment naturally, examples of such seepages being seen in the Mackenzie Delta and the intertidal zone of Alaska's North Slope. Today 80 per cent of the oil making its way into the environment is still from natural seepage. But, of course, that means 20 per cent is man-made, and that figure will inevitably rise. Natural seepage involves small quantities over large areas. The problem with man-made spills is that they are highly concentrated.

In August 1994 an oil spillage occurred in the Usinsk region of Russia's Komi Republic, south of the Arctic Circle, when twenty-three ruptures occurred in a 52km (32 mile) section of pipeline built in 1974, but not maintained during the twenty-year period since. An early attempt to contain the spill by building a dam failed, and at least 150,000 tons of oil spread across 285 acres of forest. A further 400 acres of land was stripped or destroyed by the efforts to cope with the spill. Although some oil was recovered, some was burned, the fire causing further environmental damage. Oil seeped into local waterways and in the Kolva River (and subsequently to the Pechora River) and the populations of some species of fish were reduced by 90 per cent.The clean-up operation was financed by the World Bank (1996–9), and later clean-up operations were carried out by the Lukoil company, which purchased the aging pipeline and post-Soviet facilities. The cleanup was completed in 2004. Elsewhere, studies have shown that even at twenty-eight years land at the site of diesel spills showed little recovery of vegetation and hydrocarbons remain in the soil.

It would be encouraging if such events were infrequent, but a study carried out ten years ago showed that in the Soviet/Russia oilfields almost 1,850 acres of land had been contaminated by the spillage of 130 tons of oil. Interestingly, much of the data on these spills derived from oil companies attempting to buy rival operations in post-perestroika Russia, the existence of leaking pipes allowing a bidder to reduce the price it was willing to pay.

The situation has apparently improved, but spills still occur. It would also be encouraging if it could be stated that such events were caused by poor construction or maintenance, and that these things were more common in the USSR/Russia.

In March 2006 a large oil spill was discovered at Prudhoe Bay. It transpired that the spill, the largest to date on the Alaska field and the largest in the tundra zone, involved 900 tons of oil and had been caused by corrosion of a pipeline. The worst thing was that the spill went undetected for five days before a field worker smelled crude oil while driving through the area. Subsequent inspections showed that 25km (15½ miles) of pipeline needed to be replaced because of severe corrosion. Replacement of the corroded sections of pipe caused a shutdown of the BP fields and an increase in the world price of oil. A later investigation showed that at a time when BP was making staggering annual profits, cost-cutting in pipe inspections had been instigated. The BP claim that an increase in spend would not have prevented the leaks was greeted with incredulity by many observers.

Oil spills on land are devastating for the environment, but at sea they are usually much worse, involving greater tonnages of oil, and dispersing and sinking, making the clean up more difficult and environmental damage more widespread. In March 1967 the oil tanker *Torrey Canyon* hit rocks between Cornwall and the Scilly Isles, off Britain's south-west coast. The ship had inadequate charts and radar, and apparently the cook was at the helm at the time of the disaster. Oil discharged from the tanker contaminated several hundred kilometres of French and Cornish coast, killing marine life and about 15,000 seabirds. The response to the disaster, the world's first major oil spill accident, was poor: floating booms intended to contain the oil slick proved too fragile for Atlantic winter seas and the chemical disperants employed were, in the main, used inappropriately and proved to have done more damage to the environment than the oil. In a final, desperate measure, the wrecked ship was bombed in the hope of burning the oil that remained on it. But despite the catalogue of failures, the incident did result in changes to international law, including the International Maritime Organization's 1973 International Convention for the Prevention of Pollution from Ships (as modified by the Protocol of 1978) and 1969 Civil Liability Convention (as modified by Protocols and Amendments). The latter imposed liabilities on the owners of ships without the need to prove negligence.

Twenty years after the *Torrey Canyon* hit Pollard's Rock, on 24 March 1989 the tanker *Exxon Valdez* left the Valdez oil terminal in southern Alaska, loaded with 1.3 million barrels of oil which had been pumped from the Prudhoe Bay field. A little after 12noon the ship went ashore on Bligh Reef and released 250,000 barrels (6 million gallons) into Prince William Sound. (For comparison, the *Amoco Cadiz,* released 48 million gallons in 1978, the *Torrey Canyon* released 26 million gallons in 1967, the *Brier* 19 million gallons in 1993, the *Sea Empress* 16.5 million gallons in 1996 and the *Al-Coruna* 15 million gallons in 1992.) The disaster which followed has frequently been called the world's worst environmental catastrophe. A lack of emergency preparedness, which led to a lack of immediate action, allowed a storm to disperse the oil slick. Ultimately the oil contaminated 2,500km (1,500 miles) of shoreline. The full extent of the damage to local wildlife is still debated, but about 3,000 sea otters, together with more than 200 harbour seals and 20 orca died. As many as 250,000 birds died. There was also significant damage to the marine environment. In the aftermath of the disaster, in 1994, the ExxonMobile Corporation was fined $1 billion for violations of several Acts, and a further $5 billion in punitive damages. Over the next fourteen years the Corporation fought to have the punitive damages reduced. One reason put forward as mitigation was that the Corporation had paid $2 billion on a clean-up, and $1 billion to settle private claims (neither of which, of course, would have been necessary had the tanker not run aground, and which ignored the fact that much of the clean up cost had been recovered through insurance claims). By 2007, through a succession of appeals, and some favourable rulings, the Corporation had halved the damages. Then in 2008 the US Supreme Court set aside the $2.5 billion damages on the grounds they were excessive, noting that the Corporation's position had been 'worse than negligent but less than malicious'. The Court referred the case back to a lower court, noting that punitive damages should be no higher than $507.5 million. The Corporation's position is that they should not exceed $25 million.

The Corporation has funded several hundred studies which indicate no long-term effects to the environment, though some bird species have not recovered their numbers, and in early 2007 an independent study of Prince William Sound and the Gulf of Alaska found that the oil was still a threat to wildlife as its decay was much slower than had been anticipated. The study was released the day before ExxonMobil revealed that its 2006 profit had been $39.5 billion dollars, the largest annual profit ever posted by a US company. The Corporation's profit for 2007 was $40.6 billion.

Although currently the northern oilfields are more closely scrutinized (and generally cleaner as a consequence) concerns over oil spills will increase as summer sea ice shrinks, and the North-West Passage becomes an economic route. The Passage represents a short-cut from Europe to Asia of more than 6,000km (3,750 miles) and, as has been noted with growing concern in Canada, its availability is sure to encourage some ships to attempt it. But the ice currents of the Arctic Ocean, together with local effects of climate – which, as the world is already seeing, may involve more extreme events – are likely, perhaps for many years, to create local ice conditions which catch out the unwary or ill-prepared. Unfortunately, history also shows that shipping is not always under the control of responsible owners.

Eventually a single-hulled or non-ice-strengthened, loaded tanker will hit rocks which the ice had formerly hidden, or will hit an iceberg. Or the ice will take hold of a ship. Ice has a relentless power, and will crush a ship or shift it aground. In an area where the response time to a disaster site will inevitably be much extended over that for the *Exxon Valdez*, and with a considerably more fragile ecology, the disaster will be very much worse, particularly as oil also persists for longer in the Arctic than in more southerly areas, since it may become trapped under sea ice. Oil spills on sea ice could be even more devastating. Black oil will absorb heat from the sun, causing local melting of the ice (an effect which is seen, on a much smaller scale, if seaweed is trapped in sea ice, the dark seaweed absorbing heat and creating a local melt pool) which would allow the oil to disperse below the ice.

Before considering the political implications of the search for oil and gas in the Arctic, one other energy source must be mentioned, a source which, if the scientific estimates are correct, may dwarf the area's fossil fuel deposits, but whose exploitation may hasten an already alarming climate crisis. As well as gas in its 'natural' state, gas also exists as hydrates, gas molecules encased in ice, the gas usually being methane. Gas hydrates are found in ocean or inland sea or lake sediments, at depths greater than about 300m (1,000ft) where sediment temperatures are below freezing. In the Arctic they are found within or below the permafrost and are generally concentrated where there are oil and gas reserves. Hydrates may exceed natural gas reserves by a factor of ten in energy terms, but they bring sizeable problems. One is that methane is itself a greenhouse gas and so drilling for hydrates, or even near hydrate concentrations, may cause temperature rises and pressure decreases (hydrates are at both low temperature and high pressure), and so release the gas before it can be captured, though some experts doubt that sufficiently high temperatures could be reached. There is some evidence that seafloor landslides might result when gas trapped in hydrate form is released; it has been suggested that large-scale release of methane as a result of landslides may have been the reason why the Earth warmed abruptly at the end of the last ice age. The theory suggests that reduced ocean depth resulted in the depressurization of hydrate concentrations, followed by the release of methane and a 'greenhouse' effect. Drilling could itself precipitate landslides and a potentially massive methane release. Methane release from hydrates is the likely source of the sea ice gas flares reported during 2008. If that suggestion is true, then methane release from permafrost may have already preceded hydrate exploitation, and as methane is a much more potent greenhouse gas than carbon dioxide, may result in a much faster increase in the Earth's mean temperature, and an increasingly rapid loss of Arctic sea ice. In a document published after the Arctic Inuit International Conference of 1992 (*Principles and Elements for a Comprehensive Arctic Policy*) the Conference noted that 'renewable resources must be managed and protected in a manner that maintains ecological balance, respects Inuit resource rights, and sustains the renewable resource needs of Inuit, both now and in the future.' Ironically, this has largely happened, with whaling, sealing and bear hunting severely limited and maintained by quotas, so that when the species were finally freed from the devastation of that most rapacious of hunters – man – it was just in time for them to be forced to the edge of extinction by climate change. In a terrible irony, the resource which the developed nations are now most anxious to exploit are fossil fuel reserves: seeking out the very thing whose use will enhance the warming that has allowed its discovery.

In August 2007 during a Russian expedition, carried out within the framework of the International Polar Year, two mini-submersibles (one of which was operated by a Russian, but included two tourists, a Swede and an Australian, who paid a stiff price for the privilege, and whose presence made the dive financially feasible) reached the sea floor under the North Pole (at a depth of 4,261m/14,000ft) for the first time, and planted a flag. The planting of the flag, televised in Russia as it represented a return to the Arctic after the lean post-perestroika period, was seen by the international community as staking a territorial claim. The flag-planting was greeted with particular alarm in the United Sates, where there was an immediate rush to point out American precedence over the Pole, noting that Robert Peary had been the first to reach it. One report, perhaps recalling that Peary's claim is dismissed by many polar experts, noted that a US nuclear submarine had genuinely been there ahead of the Russians (the *Skate* in 1959), apparently forgetting that a Soviet aircraft had landed at the Pole in 1948. The concern was that the Russian flag-planting was linked to its claim that the Lomonosov Ridge was an extension of Russia's northern continental shelf and, as such, supported a claim to a greater share of the Arctic than the standard sector claim. Russia had made a formal claim regarding the Lomonosov Ridge to the United Nations Commission on the Limits of the Continental Shelf in 2001, but the Commission was far from convinced, asking for further seismic and sonar measurements, which the Russians have been collecting. The claim has been strongly contested by Denmark, Norway and the United States and especially Canada, which is making a similar claim after further survey work on the Ridge. Peter MacKay, the former Canadian Foreign Minister stated: 'This is the true north strong and free, and they [the Russians] are fooling themselves if they think dropping a flag on the ocean floor is going to change anything. There is no question over Canadian sovereignty in the Arctic. We've made that very clear. We've established – a long time ago – that these are Canadian waters and this is Canadian property. You can't go around the world these days dropping a flag somewhere. This isn't the fourteenth or fifteenth century.' MacKay's remarks prompted an immediate response from Russian Minister Sergey Lavrov who said: 'I read reports of the statements made by my Canadian colleague, Peter MacKay. I know him quite well – it's very unlike him. I was sincerely astonished by "flag planting." No one engages in flag planting. When pioneers reach a point hitherto unexplored by anybody, it is customary to leave flags there. Such was the case on the Moon. As to the legal aspect of the matter, we from the outset said that this expedition was part of the work being carried out under the UN Convention on the Law of the Sea, where Russia's claim to submerged ridges, which we believe to be an extension of our shelf, is being considered. We know that this has to be proved. The ground samples that were taken will serve the work to prepare that evidence.'

Russia's claim to the Lomonosov Ridge has also been received sceptically by geologists in non-Arctic nations, who point out that the Ridge is actually a tectonic spreading ridge and so not associated with any land mass. However, the current paradigm of continental drift postulates that the mid-ocean ridges are the birthplace of the sea floor, which travels to the sides and thus pushes the continents apart. Denmark has recently indicated that it is considering its own claim to the Ridge, a claim which is likely to be equally strongly contested. The point of the rival claims is that while sovereignty claims over land masses depend (to a lesser or greater extent depending upon an interpretation of international law) on occupation, there is no such requirement for a continental shelf. A nation has the right to develop its continental shelf without even the necessity to formally proclaim its sovereignty over it. With their resource-based economies, it is no surprise that the Arctic nations are keen to develop their northern seas, reducing their dependencies on imported energy (in the case of Denmark, Canada and the United States) or increasing the influence that control of energy supplies brings with it.

In 2007 the Canadian Government declared its intention to build new ice-breakers and to develop a deep water port on Baffin Island. The details of the plan, which a further announcement in August 2008 confirmed, now appear fixed, the ship having a name (it will be named after the late Prime Minister John Diefenbaker) and a $720 million budget. Iqaluit and Nanisivik (both on Baffin Island) have been suggested as sites for the port. Iqaluit has the advantages of having the only metalled runway in Nunavut – Nanisivik has a gravel strip – and of being larger, with a better developed infrastructure, and has the kudos of being the capital of Nunavut. Its disadvantage is its southern position, Nanisivik being in Admiralty Inlet at the northern end of the island, and therefore close to Lancaster Sound, the entrance to all realistic versions of a North-West Passage. Exactly what the ice-breakers will look like, and how many there will be seems likely to be defined by cost. One suggestion in the 2007 announcement was for a small fleet of heavy coastguard vessels, these rapidly being termed 'slush-breakers' by those who considered the Government's proposal too timid. But the 2008 announcement was more defiant, stating that the intention was to build one or more armoured heavy ice-breakers, though the weaponry was not defined. The proposed build has its critics, but supporters and critics alike are concerned over the cost of the programme.

Nominally, the Canadian rationale for the ships would be patrol of the North-West Passage, reinforcing the country's claim to sovereignty over the waters between the islands of its Arctic archipelago, but it is easy to see in the announcement an uneasy concern over the Russian interest in the Arctic Basin. In Russia there was a swift reaction to the Canadian plans to build ice-breakers. Having completed the *50 Let Pobedy* (see below) – building started in 1989; it was completed in 2007 – the country had undergone a period of marked euphoria about financial potential of Russian economy before the global financial crisis of late-2008. In October ITAR-TASS (the government news agency) announced that Russia was going to build military ice-breakers capable of performing coastal guard functions ('slush-breakers' in Canadian terminology). The military use of ice-breakers by Russia has a long history: in 1942 the *Alexander Sibiriakov* was sunk in a battle with German Pocket battleship *Admiral Scheer*. Ice-breaker class coastal guard ships (similar to the Canadian Amundsen class ships) were built in Russia during 1975–81, six are currently still in service. The announcement may have indicated a desire to commission new ships to replace these existing ships, but with the onset of the global financial crisis, no further announcements on the ships has been made.

Russia already has a number of nuclear-powered ice-breaker, each with a greater capability in thick ice than anything the United States has, or the Canadians will have. The first of Russia's fleet was the *Lenin*, launched in 1959 and decommissioned in 1989. The *Arktika*, launched in 1975, was the first surface vessel to reach the North Pole: the ship was decommissioned in October 2008 and is now being dismantled. The ship was the first of the so-called Arktika class of ice-breakers which include the *Sibir* (no longer operational), *Rossiya, Sovetskiy Soyuz, Yamal* and *50 Years since* (or *of*) *Victory*. The last named (in Russian *50 лет Победы*, pronounced *50 Let Pobedy*) was not completed until 2007. In 2009 it will be used as part of the American-based tour company Quark Expeditions' series of Arctic voyages, taking around 120 passengers to the North Pole from Murmansk. The Arktika class ships have two nuclear reactors, each providing about 170MW of power allowing them to crash through ice up to 3m (10ft) thick and to achieve a maximum speed in ice of about 10 knots, but of more than 20 knots in open water. In addition, the Russians have two Taimyr class nuclear ice-breakers (the *Taimyr* and the *Vaigach*), which are powered by two 135MW reactors giving a top speed of about 18 knots and an ice capability to at least 2m (7ft). The two ships are primarily used to break river ice, particularly on the Yenisey River, allowing cargo ships to travel between Dickson and Noril'sk. The *Seymorput* is a nuclear-powered container ship used on the Northern Sea Route; there are now plans to convert it into a first nuclear-powered oil-drilling ship in the world. In 2008 Russia declared its intention to construct a new generation of nuclear-powered ice-breakers, the first to be available by 2015. The possibility of a fleet of military nuclear-powered ice-breakers operating deep in the Arctic Basin concerns the other Arctic nations. The Russia of Vladimir Putin (no longer the President, but still, it would appear, the man in charge) looks increasingly authoritarian, bearing some striking similarities to the Soviet past. There are some who believe that the Russian aim is to add Arctic fuel supplies to its massive land reserves. Gazprom already controls 25 per cent of European gas supplies. If Russia controlled Arctic oil and gas its power to influence world events would be strengthened immeasurably. One area of contention in Arctic waters is the border between Russia and the US in the Bering Sea, particularly as current indications are that fuel reserves are grouped on the American side. At present the most powerful of US ice-breakers has to back up and ram ice about 2m (7ft) thick which *50 Let Pobedy* can plough through at more than 10 knots. The Bering Sea is no longer routinely ice-bound, but the Chukchi Sea is, and so are the waters of the Arctic Basin.

In August 2007 Russia resumed regular air force patrols over the Arctic Ocean, the planes including both bombers, refuelled in the air to increase their range, and anti-submarine aircraft. For the first time since the end of the Cold War flights have occasionally penetrated the US identification zone extending from Alaska. At the same time, the Russian navy has increased its presence in Arctic waters, particularly with anti-submarine destroyers and missile cruisers, while the Russian Defence Minister noted that Russia had military units spread across its northern coast which had been trained in Arctic warfare.

But, in May 2008 representatives from Canada, Denmark, Norway, Russia and the United States met in Ilulissat in west Greenland to decide on how ownership of the Arctic Basin should be apportioned. Excluded from the meeting were the other three Arctic nations – Finland, Iceland and Sweden – on the grounds that they have no Arctic coastline, and any representative from the native peoples of the Arctic. None of the five came with any claim to the Basin, and all five reaffirmed their view that existing international treaties were the correct basis for negotiating the Basin's future. One such treaty would be UNCLOS, as the meeting agreed (despite the fact that the United States had not ratified the Convention). The Russian Foreign Minister said he hoped the peaceful outcome of the meeting would reassure those who felt that a 'battle for the Arctic Ocean' was likely to take place.

At much the same time as the Ilulissat meeting, the price of oil rose to $147 per barrel. The financial crisis which followed saw the price fall below $40, despite OPEC's strenuous efforts to prevent the collapse. The price reduction was welcomed, not only by those who had heating and transport bills to pay, but by governments pleased to see a limitation to OPEC's power. The price reduction also made Arctic oil and gas exploration less attractive. But the price of a barrel will inevitably rise again and eyes will turn north once more. Then the scramble for the Arctic will result in a new, and far more dangerous, Cold War.

During the Cold War it was said that the belligerent posturing of the USA and the USSR over their respective idealogies and spheres of influence was similar to two fleas fighting over which owned the dog. The conflict over Arctic ownership and exploitation alters the analogy. The dog is now dying from an infection carried by the fleas. And there isn't another dog.

Appendix 1
The Svalbard Treaty (1920)

Treaty between Norway, The United States of America, Denmark, France, Italy, Japan, the Netherlands, Great Britain and Ireland and the British overseas Dominions and Sweden concerning Spitsbergen signed in Paris, 9 February 1920.

The President of The United States of America; His Majesty the King of Great Britain and Ireland and of the British Dominions beyond the Seas, Emperor of India; His Majesty the King of Denmark; the President of the French Republic; His Majesty the King of Italy; His Majesty the Emperor of Japan; His Majesty the King of Norway; Her Majesty the Queen of the Netherlands; His Majesty the King of Sweden,

Desirous, while recognising the sovereignty of Norway over the Archipelago of Spitsbergen, including Bear Island, of seeing these territories provided with an equitable regime, in order to assure their development and peaceful utilisation,

Have appointed as their respective Plenipotentiaries with a view to concluding a Treaty to this effect:

The President of the United States of America: Mr. Hugh Campbell Wallace, Ambassador Extraordinary and Plenipotentiary of the United States of America at Paris;

His Majesty the King of Great Britain and Ireland and of the British Dominions beyond the Seas, Emperor of India: The Right Honourable the Earl of Derby, K.G., G.C.V.O., C.B., His Ambassador Extraordinary and Plenipotentiary at Paris;

and for the *Dominion of Canada:* The Right Honourable Sir George Halsey Perley, K.C.M.G., High Commissioner for Canada in the United Kingdom;

for the *Commonwealth of Australia:* The Right Honourable Andrew Fisher, High Commissioner for Australia in the United Kingdom;

for the *Dominion of New Zealand:* The Right Honourable Sir Thomas MacKenzie, K.C.M.G., High Commissioner for New Zealand in the United Kingdom;

for the *Union of South Africa:* Mr. Reginald Andrew Blankenberg, O.B.E, Acting High Commissioner for South Africa in the United Kingdom;

for *India:* The. Right Honourable the Earl of Derby, K.G., G.C.V.O., C. B.;

His Majesty the King of Denmark: Mr. Herman Anker Bernhoft, Envoy Extraordinary and Minister Plenipotentiary of H.M. the King of Denmark at Paris;

President of the French Republic: Mr. Alexandra Millerand, President of the Council, Minister for Foreign Affairs;

His Majesty the King of Italy: The Honourable Maggiorino Ferraris, Senator of the Kingdom;

His Majesty the Emperor of Japan: Mr. K. Matsui, Ambassador Extraordinary and Plenipotentiary of H.M. the Emperor of Japan at Paris;

His Majesty the King of Norway: Baron Wedel Jarlsberg, Envoy Extraordinary and Minister Plenipotentiary of H.M. the King of Norway at Paris;

Her Majesty the Queen of *the Netherlands:* Mr. John London, Envoy Extraordinary and Minister Plenipotentiary of H.M. the Queen of the Netherlands at Paris;

His Majesty the King of Sweden: Count J.-J.-A. Ehrensvärd, Envoy Extraordinary and Minister Plenipotentiary of H.M. the King of Sweden at Paris;

Who, having communicated their full powers, found in good and due form, have agreed as follows:

Article 1.
The High Contracting Parties undertake to recognise, subject to the stipulations of the present Treaty, the full and absolute sovereignty of Norway over the Archipelago of Spitsbergen, comprising, with Bear Island or Beeren-Eiland [i.e. Bjørnøya], all the islands situated between 10° and 35° longitude East of Greenwich and between 74° and 81° latitude North, especially West Spitsbergen, North-East Land [i.e. Nordaustlandet], Barents Island [i.e. Barentsøya], Edge Island [Edgeøya], Wiche Islands [i.e. Kong Karls Land], Hope Island or Hopen-Eiland [i.e. Hopen], and Prince Charles Foreland [Prins Karls Forland], together with all islands great or small and rocks appertaining thereto (Author note: a map of the archipelago was attached to the Treaty).

Article 2.
Ships and nationals of all the High Contracting Parties shall enjoy equally the rights of fishing and hunting in the territories specified in Article I and in their territorial waters.

Norway shall be free to maintain, take or decree suitable measures to ensure the preservation and, if necessary, the reconstitution of the fauna and flora of the said regions, and their territorial waters; it being clearly understood that these measures shall always be applicable equally to the nationals of all the High Contracting Parties without any exemption, privilege or favour whatsoever, direct or indirect to the advantage of any one of them.

Occupiers of land whose rights have been recognised in accordance with the terms of Articles 6 and 7 will enjoy the exclusive right of hunting on their own land: (I) in the neighbourhood of their habitations, houses, stores, factories and installations, constructed for the purpose of developing their property, under conditions laid down by the local police regulations; (2) within a radius of 10 kilometres round the headquarters of their place of business or works; and in both cases, subject always to the observance of regulations made by the Norwegian Government in accordance with the conditions laid down in the present Article.

Article 3.
The nationals of all the High Contracting Parties shall have equal liberty of access and entry for any reason or object whatever to the waters, fjords and ports of the territories specified in Article I; subject to the observance of local laws and regulations, they may carry on there without impediment all maritime, industrial, mining and commercial operations on a footing of absolute equality.

They shall be admitted under the same conditions of equality to the exercise and practice of all maritime, industrial, mining or commercial enterprises both on land and in the territorial waters, and no monopoly shall be established on any account or for any enterprise whatever.

Notwithstanding any rules relating to coasting trade which may be in force in Norway, ships of the High Contracting Parties going to or coming from the territories specified in Article I shall have the right to put into Norwegian ports on their outward or homeward voyage for the purpose of taking on board or disembarking passengers or cargo going to or coming from the said territories, or for any other purpose.

It is agreed that in every respect and especially with regard to exports, imports and transit traffic, the nationals of all the High Contracting Parties, their ships and goods shall not be subject to any charges or restrictions whatever which are not borne by the nationals, ships or goods which enjoy in Norway the treatment of the most favoured nation; Norwegian nationals, ships or goods being for this purpose assimilated to those of the other High Contracting Parties, and not treated more favourably in any respect.

No charge or restriction shall be imposed on the exportation of any goods to the territories of any of the Contracting Powers other or more onerous than on the exportation of similar goods to the territory of any other Contracting Power (including Norway) or to any other destination.

Article 4.
All public wireless telegraphy stations established or to be established by, or with the authorisation of, the Norwegian Government within the territories referred to in Article 1 shall always be open on a footing of absolute equality to communications from ships of all flags and from nationals of the High Contracting Parties, under the conditions laid down in the Wireless Telegraphy Convention of July 5, 1912, or in the subsequent International Convention which may be concluded to replace it.

Subject to international obligations arising out of a state of war, owners of landed property shall always be at liberty to establish and use for their own purposes wireless telegraphy installations, which shall be free to communicate on private business with fixed or moving wireless stations, including those on board ships and aircraft.

Article 5.
The High Contracting Parties recognise the utility of establishing an international meteorological station in the territories specified in Article 1, the organisation of which shall form the subject of a subsequent Convention.

Conventions shall also be concluded laying down the conditions under which scientific investigations may be conducted in the said territories.

Article 6.
Subject to the provisions of the present Article, acquired rights of nationals of the High Contracting Parties shall be recognised.

Claims arising from taking possession or from occupation of land before the signature of the present Treaty shall be dealt with in accordance with the Annex hereto, which will have the same force and effect as the present Treaty.

Article 7.
With regard to methods of acquisition, enjoyment and exercise of the right of ownership of property, including mineral rights, in the territories specified in Article 1, Norway undertakes to grant to all nationals of the High Contracting Parties treatment based on complete equality and in conformity with the stipulations of the present Treaty.

Expropriation may be resorted to only on grounds of public utility and on payment of proper compensation.

Article 8.
Norway undertakes to provide for the territories specified in Article 1 mining regulations which, especially from the point of view of imposts, taxes or charges of any kind, and of general or particular labour conditions, shall exclude all privileges, monopolies or favours for the benefit of the State or of the nationals of any one of the High Contracting Parties, including Norway, and shall guarantee to the paid staff of all categories the remuneration and protection necessary for their physical, moral and intellectual welfare.

Taxes, dues and duties levied shall be devoted exclusively to the said territories and shall not exceed what is required for the object in view.

So far, particularly, as the exportation of minerals is concerned, the Norwegian Government shall have the right to levy an export duty which shall not exceed 1 per cent of the maximum value of the minerals exported up to 100,000 tons, and beyond that quantity the duty will be proportionately diminished. The value shall be fixed at the end of the navigation season by calculating the average free on board price obtained.

Three months before the date fixed for their coming into force, the draft mining regulations shall be communicated by the Norwegian Government to the other Contracting Powers. If during this period one or more of the said Powers propose to modify these regulations before they are applied, such proposals shall be communicated by the Norwegian Government to the other Contracting Powers in order that they may be submitted to examination and the decision of a Commission composed of one representative of each of the said Powers. This Commission shall meet at the invitation of the Norwegian Government and shall come to a decision within a period of three months from the date of its first meeting. Its decisions shall be taken by a majority.

Article 9.
Subject to the rights and duties resulting from the admission of Norway to the League of Nations, Norway undertakes not to create nor to allow the establishment of any naval base in the territories specified in Article 1 and not to construct any fortification in the said territories, which may never be used for warlike purposes.

Article 10.
Until the recognition by the High Contracting Parties of a Russian Government shall permit Russia to adhere to the present Treaty, Russian nationals and companies shall enjoy the same rights as nationals of the High Contracting Parties.

Claims in the territories specified in Article 1 which they may have to put forward shall be presented under the conditions laid down in the present Treaty (Article 6 and Annex) through the intermediary of the Danish Government, who declare their willingness to lend their good offices for this purpose.

The present Treaty, of which the French and English texts are both authentic, shall be ratified.

Ratifications shall be deposited at Paris as soon as possible.

Powers of which the seat of the Government is outside Europe may confine their action to informing the Government of the French Republic, through their diplomatic representative at Paris, that their ratification has been given, and in this case, they shall transmit the instrument as soon as possible.

The present Treaty will come into force, in so far as the stipulations of Article 8 are concerned, from the date of its ratification by all the signatory Powers; and in all other respects on the same date as the mining regulations provided for in that Article.

Third Powers will be invited by the Government of the French Republic to adhere to the present Treaty duly ratified. This adhesion shall be effected by a communication addressed to the French Government, which will undertake to notify the other Contracting Parties.

In witness whereof the above named Plenipotentiaries have signed the present Treaty.

Done at Paris, the ninth day of February, 1920, in duplicate, one copy to be transmitted to the Government of His Majesty the King of Norway, and one deposited in the archives of the French Republic; authenticated copies will be transmitted to the other Signatory Powers.

Annex

I.

(I) Within three months from the coming into force of the present Treaty, notification of all claims to land which had been made to any Government before the signature of the present Treaty must be sent by the Government of the claimant to a Commissioner charged to examine such claims. The Commissioner will be a judge or jurisconsul of Danish nationality possessing the necessary qualifications for the task, and shall be nominated by the Danish Government.

(2) The notification must include a precise delimitation of the land claimed and be accompanied by a map on a scale of not less than I/I,000,000 on which the land claimed is clearly marked.

(3) The notification must be accompanied by the deposit of a sum of one penny for each acre (40 acres) of land claimed, to defray the expenses of the examination of the claims.

(4) The Commissioner will be entitled to require from the claimants any further documents or information which he may consider necessary.

(5) The Commissioner will examine the claims so notified. For this purpose he will be entitled to avail himself such expert assistance as he may consider necessary, and in case of need to cause investigations to be carried out on the spot.

(6) The remuneration of the Commissioner will be fixed by agreement between the Danish Government and the other Governments concerned. The Commissioner will fix the remuneration of such assistants as he considers it necessary to employ.

(7) The Commissioner, after examining the claims, will prepare a report showing precisely the claims which he is of opinion should be recognised at once and those which, either because they are disputed or for any other reason, he is of opinion should be submitted to arbitration as hereinafter provided. Copies of this report will be forwarded by Commissioner to the Governments concerned.

(8) If the amount of the sums deposited in accordance with clause (3) is insufficient to cover the expenses of the examination of the claims, the Commissioner will, in every case where he is of opinion that a claim should be recognised, at once state what further sum the claiment should be required to pay. This sum will be based on the amount of the land to which the claimant's title is recognised.

If the sums deposited in accordance with clause (3) exceed the expenses of the examination, the balance will devoted to the cost of the arbitration hereinafter provided for.

(9) Within three months from the date of the report referred to in clause (7) of this paragraph, the Norwegian Government shall take the necessary steps to confer upon claimants whose claims have been recognised by the Commissioner a valid title securing to them the exclusive property in the land in question, in accordance with the laws and regulations in force or to be enforced in the territories specified in Article I of the present Treaty, and subject to the mining regulations referred to in Article 8 of the present Treaty.

In the event, however, of a further payment being required in accordance with clause (8) of this paragraph, a provisional title only will be delivered, which title will become definitive on payment by the claimant, within such reasonable period as the Norwegian Government may fix, of the further sum required of him.

2.

Claims which for any reason the Commissioner referred to in clause (I) of the preceding paragraph has not recognised as valid will be settled in accordance with the following provisions:

(1) Within three months from the date of the report referred to in clause (7) of the preceding paragraph, each of the Governments whose nationals have been found to possess claims which have not been recognised will appoint an arbitrator.

The Commissioner will be the President of the Tribunal so constituted. In cases of equal division of opinion, he shall have the deciding vote. He will nominate a Secretary to receive the documents referred to in clause (2) of this paragraph and to make the necessary arrangements for the meeting of the Tribunal.

(2) Within one month from the appointment of the Secretary referred to in clause (1) the claimants concerned will send to him through the intermediary of their respective Governments statements indicating precisely their claims and accompanied by such documents and arguments as they may wish to submit in support thereof.

(3) Within two months from the appointment of the Secretary referred to in clause (1) the Tribunal shall meet at Copenhagen for the purpose of dealing with the claims which have been submitted to it.

(4) The language of the Tribunal shall be English. Documents or arguments may be submitted to it by the interested parties in their own language, but in that case must be accompanied by an English translation.

(5) The claimants shall be entitled, if they so desire, to be heard by the Tribunal either in person or by counsel, and the Tribunal shall be entitled to call upon the claimants to present such additional explanations, documents or arguments as it may think necessary.

(6) Before the hearing of any case the Tribunal shall require from the parties a deposit or security for such sum as it may think necessary to cover the share of each party in the expenses of the Tribunal. In fixing the amount of such sum the Tribunal shall base itself principally on the extent of the land claimed. The Tribunal shall also have power to demand a further deposit from the parties in cases where special expense is involved.

(7) The honorarium of the arbitrators shall be calculated per month, and fixed by the Governments concerned. The salary of the Secretary and any other persons employed by the Tribunal shall be fixed by the President.

(8) Subject to the provisions of this Annex the Tribunal shall have full power to regulate its own procedure.

(9) In dealing with the claims the Tribunal shall take into consideration:

(a) any applicable rules of International Law;

(b) the general principles of justice and equity;

(c) the following circumstances:

(i) the date on which the land claimed was first occupied by the claimant or his predecessors in title;

(ii) the date on which the claim was notified to the Government of the claimant;

(iii) the extent to which the claimant or his predecessors in title have developed and exploited the land claimed. In this connection the Tribunal shall take into account the extent to which the claimants may have been prevented from developing their undertakings by conditions or restrictions resulting from the war of 1914-1919.

(10) All the expenses of the Tribunal shall be divided among the claimants in such proportion as the Tribunal shall decide. If the amount of the sums paid in accordance with clause (6) is larger than the expenses of the Tribunal, the balance shall be returned to the parties

whose claims have been recognised in such proportion as the Tribunal shall think fit.
(II) The decisions of the Tribunal shall be communicated by it to the Governments concerned, including in every case the Norwegian Government.

The Norwegian Government shall within three months from the receipt of each decision take the necessary steps to confer upon the claimant whose claims have been recognised by the Tribunal valid title to the land in question, in accordance with the laws and regulations in force or the be enforced in the territories specified in Article I, and subject to the mining regulations referred to in Article 8 of the present Treaty. Nevertheless, the titles so conferred will only become definitive on the payment by the claimant concerned, within such reasonable period as the Norwegian Government may fix, of his share of the expenses of the Tribunal.

3.
Any claims which are not notified to the Commissioner in accordance with clause (I) of paragraph I, or which not having been recognised by him are not submitted to the Tribunal in accordance with paragraph 2, will be finally extinguished.

Appendix 2
The Antarctic Treaty (1959)

Preamble
The Governments of Argentina, Australia, Belgium, Chile, the French Republic, Japan, New Zealand, Norway, the Union of South Africa, the Union of Soviet Socialist Republics, the United Kingdom of Great Britain and Northern Ireland, and the United States of America,

Recognizing that it is in the interest of all mankind that Antarctica shall continue for ever to be used exclusively for peaceful purposes and shall not become the scene or object of international discord;

Acknowledging the substantial contributions to scientific knowledge resulting from international cooperation in scientific investigation in Antarctica;

Convinced that the establishment of a firm foundation for the continuation and development of such cooperation on the basis of freedom of scientific investigation in Antarctica as applied during the International Geophysical Year accords with the interests of science and the progress of all mankind;

Convinced also that a treaty ensuring the use of Antarctica for peaceful purposes only and the continuance of international harmony in Antarctica will further the purposes and principles embodied in the Charter of the United Nations;

Have agreed as follows:

Article I – Peaceful purposes
1. Antarctica shall be used for peaceful purposes only. There shall be prohibited, inter alia, any measure of a military nature, such as the establishment of military bases and fortifications, the carrying out of military manoeuvres, as well as the testing of any type of weapon.
2. The present Treaty shall not prevent the use of military personnel or equipment for scientific research or for any other peaceful purpose.

Article II – Freedom of scientific investigation
Freedom of scientific investigation in Antarctica and cooperation toward that end, as applied during the International Geophysical Year, shall continue, subject to the provisions of the present Treaty.

Article III – International scientific cooperation
1. In order to promote international cooperation in scientific investigation in Antarctica, as provided for in Article II of the present Treaty, the Contracting Parties agree that, to the greatest extent feasible and practicable:

a. information regarding plans for scientific programs in Antarctica shall be exchanged to permit maximum economy of and efficiency of operations;
b. scientific personnel shall be exchanged in Antarctica between expeditions and stations;
c. scientific observations and results from Antarctica shall be exchanged and made freely available.

Article IV – Territorial sovereignty
1. Nothing contained in the present Treaty shall be interpreted as:

a. a renunciation by any Contracting Party of previously asserted rights of or claims to territorial sovereignty in Antarctica;
b. a renunciation or diminution by any Contracting Party of any basis of claim to territorial sovereignty in Antarctica which it may have whether as a result of its activities or those of its nationals in Antarctica, or otherwise;

c. prejudicing the position of any Contracting Party as regards its recognition or non-recognition of any other State,s rights of or claim or basis of claim to territorial sovereignty in Antarctica.

2. No acts or activities taking place while the present Treaty is in force shall constitute a basis for asserting, supporting or denying a claim to territorial sovereignty in Antarctica or create any rights of sovereignty in Antarctica. No new claim, or enlargement of an existing claim, to territorial sovereignty in Antarctica shall be asserted while the present Treaty is in force.

Article V – Nuclear activity

1. Any nuclear explosions in Antarctica and the disposal there of radioactive waste material shall be prohibited.

2. In the event of the conclusion of international agreements concerning the use of nuclear energy, including nuclear explosions and the disposal of radioactive waste material, to which all of the Contracting Parties whose representatives are entitled to participate in the meetings provided for under Article IX are parties, the rules established under such agreements shall apply in Antarctica.

Article VI – Geographical coverage

The provisions of the present Treaty shall apply to the area south of 60° South Latitude, including all ice shelves, but nothing in the present Treaty shall prejudice or in any way affect the rights, or the exercise of the rights, of any State under international law with regard to the high seas within that area.

Article VII – Inspections

1. In order to promote the objectives and ensure the observance of the provisions of the present Treaty, each Contracting Party whose representatives are entitled to participate in the meetings referred to in Article IX of the Treaty shall have the right to designate observers to carry out any inspection provided for by the present Article. Observers shall be nationals of the Contracting Parties which designate them. The names of observers shall be communicated to every other Contracting Party having the right to designate observers, and like notice shall be given of the termination of their appointment.

2. Each observer designated in accordance with the provisions of paragraph 1 of this Article shall have complete freedom of access at any time to any or all areas of Antarctica.

3. All areas of Antarctica, including all stations, installations and equipment within those areas, and all ships and aircraft at points of discharging or embarking cargoes or personnel in Antarctica, shall be open at all times to inspection by any observers designated in accordance with paragraph 1 of this Article.

4. Aerial observation may be carried out at any time over any or all areas of Antarctica by any of the Contracting Parties having the right to designate observers.

5. Each Contracting Party shall, at the time when the present Treaty enters into force for it, inform the other Contracting Parties, and thereafter shall give them notice in advance, of

a. all expeditions to and within Antarctica, on the part of its ships or nationals, and all expeditions to Antarctica organized in or proceeding from its territory;

b. all stations in Antarctica occupied by its nationals; and

c. any military personnel or equipment intended to be introduced by it into Antarctica subject to the conditions prescribed in paragraph 2 of Article I of the present Treaty.

Article VIII – Jurisdiction

1. In order to facilitate the exercise of their functions under the present Treaty, and without prejudice to the respective positions of the Contracting Parties relating

to jurisdiction over all other persons in Antarctica, observers designated under paragraph I of Article VII and scientific personnel exchanged under sub-paragraph I(b) of Article III of the Treaty, and members of the staffs accompanying any such persons, shall be subject only to the jurisdiction of the Contracting Party of which they are nationals in respect of all acts or omissions occurring while they are in Antarctica for the purpose of exercising their functions.
2. Without prejudice to the provisions of paragraph I of this Article, and pending the adoption of measures in pursuance of subparagraph I(e) of Article IX, the Contracting Parties concerned in any case of dispute with regard to the exercise of jurisdiction in Antarctica shall immediately consult together with a view to reaching a mutually acceptable solution.

Article IX – Treaty Meetings
I. Representatives of the Contracting Parties named in the preamble to the present Treaty shall meet at the City of Canberra within two months after the date of entry into force of the Treaty, and thereafter at suitable intervals and places, for the purpose of exchanging information, consulting together on matters of common interest pertaining to Antarctica, and formulating and considering, and recommending to their Governments, measures in furtherance of the principles and objectives of the Treaty, including measures regarding:

a. use of Antarctica for peaceful purposes only;
b. facilitation of scientific research in Antarctica;
c. facilitation of international scientific cooperation in Antarctica;
d. facilitation of the exercise of the rights of inspection provided for in Article VII of the Treaty;
e. questions relating to the exercise of jurisdiction in Antarctica;
f. preservation and conservation of living resources in Antarctica.

2. Each Contracting Party which has become a party to the present Treaty by accession under Article XIII shall be entitled to appoint representatives to participate in the meetings referred to in paragraph I of the present Article, during such times as that Contracting Party demonstrates its interest in Antarctica by conducting substantial research activity there, such as the establishment of a scientific station or the despatch of a scientific expedition.
3. Reports from the observers referred to in Article VII of the present Treaty shall be transmitted to the representatives of the Contracting Parties participating in the meetings referred to in paragraph I of the present Article.
4. The measures referred to in paragraph I of this Article shall become effective when approved by all the Contracting Parties whose representatives were entitled to participate in the meetings held to consider those measures.
5. Any or all of the rights established in the present Treaty may be exercised as from the date of entry into force of the Treaty whether or not any measures facilitating the exercise of such rights have been proposed, considered or approved as provided in this Article.

Article X – Activities contrary to Treaty
Each of the Contracting Parties undertakes to exert appropriate efforts, consistent with the Charter of the United Nations, to the end that no one engages in any activity in Antarctica contrary to the principles or purposes of the present Treaty.

Article XI – Disputes between Parties
I. If any dispute arises between two or more of the Contracting Parties concerning the interpretation or application of the present Treaty, those Contracting Parties shall consult among themselves with a view to having the dispute resolved by negotiation,

inquiry, mediation, conciliation, arbitration, judicial settlement or other peaceful means of their own choice.
2. Any dispute of this character not so resolved shall, with the consent, in each case, of all parties to the dispute, be referred to the International Court of Justice for settlement; but failure to reach agreement on reference to the International Court shall not absolve parties to the dispute from the responsibility of continuing to seek to resolve it by any of the various peaceful means referred to in paragraph I of this Article.

Article XII – Modification and duration

I.

a. The present Treaty may be modified or amended at any time by unanimous agreement of the Contracting Parties whose representatives are entitled to participate in the meetings provided for under Article IX. Any such modification or amendment shall enter into force when the depositary Government has received notice from all such Contracting Parties that they have ratified it. b. Such modification or amendment shall thereafter enter into force as to any other Contracting Party when notice of ratification by it has been received by the depositary Government. Any such Contracting Party from which no notice of ratification is received within a period of two years from the date of entry into force of the modification or amendment in accordance with the provision of subparagraph I(a) of this Article shall be deemed to have withdrawn from the present Treaty on the date of the expiration of such period.

2.

a. If after the expiration of thirty years from the date of entry into force of the present Treaty, any of the Contracting Parties whose representatives are entitled to participate in the meetings provided for under Article IX so requests by a communication addressed to the depositary Government, a Conference of all the Contracting Parties shall be held as soon as practicable to review the operation of the Treaty.
b. Any modification or amendment to the present Treaty which is approved at such a Conference by a majority of the Contracting Parties there represented, including a majority of those whose representatives are entitled to participate in the meetings provided for under Article IX, shall be communicated by the depositary Government to all Contracting Parties immediately after the termination of the Conference and shall enter into force in accordance with the provisions of paragraph I of the present Article
c. If any such modification or amendment has not entered into force in accordance with the provisions of subparagraph I(a) of this Article within a period of two years after the date of its communication to all the Contracting Parties,any Contracting Party may at any time after the expiration of that period give notice to the depositary Government of its withdrawal from the present Treaty; and such withdrawal shall take effect two years after the receipt of the notice by the depositary Government.

Article XIII – Ratification and entry into force

I. The present Treaty shall be subject to ratification by the signatory States. It shall be open for accession by any State which is a Member of the United Nations, or by any other State which may be invited to accede to the Treaty with the consent of all the Contracting Parties whose representatives are entitled to participate in the meetings provided for under Article IX of the Treaty.
2. Ratification of or accession to the present Treaty shall be effected by each State in accordance with its constitutional processes.
3. Instruments of ratification and instruments of accession shall be deposited

with the Government of the United States of America, hereby designated as the depositary Government.
4. The depositary Government shall inform all signatory and acceding States of the date of each deposit of an instrument of ratification or accession, and the date of entry into force of the Treaty and of any modification or amendment thereto.
5. Upon the deposit of instruments of ratification by all the signatory States, the present Treaty shall enter into force for those States and for States which have deposited instruments of accession. Thereafter the Treaty shall enter into force for any acceding State upon the deposit of its instruments of accession.
6. The present Treaty shall be registered by the depositary Government pursuant to Article 102 of the Charter of the United Nations.

Article XIV – Deposition
The present Treaty, done in the English, French, Russian and Spanish languages, each version being equally authentic, shall be deposited in the archives of the Government of the United States of America, which shall transmit duly certified copies thereof to the Governments of the signatory and acceding States.

Important environmental protection agreements since the signing of the Treaty include:

Agreed Measures for the Conservation of Antarctic Fauna and Flora (1964).

The Convention for the Conservation of Antarctic Seals (1972)

The Convention for the Conservation of Antarctic Marine Living Resources (1980).

And, most importantly, the Protocol on Environmental Protection to the Antartcitc Treaty, which came into force in 1998. The Protocol prohibits all activities relating to mineral resources except those related specifically to science and provides for the protection of the environment through five annexes on marine pollution, fauna and flora, environmental impact assessments, waste management, and the defining of protected areas.

Appendix 3

The International Agreement on the Conservation of Polar Bears and Their Habitat (1976)

THE GOVERNMENTS of Canada, Denmark, Norway, the Union of Soviet Socialist Republics, and the United States of America,

RECOGNISING the special responsibilities and special interests of the States of the Arctic Region in relation to the protection of the fauna and flora of the Arctic Region;

RECOGNISING that the polar bear is a significant resource of the Arctic Region which requires additional protection;

HAVING DECIDED that such protection should be achieved through coordinated nation measures taken by the States of the Arctic Region;

DESIRING to take immediate action to bring further conservation and management measures into effect;

HAVE AGREED AS FOLLOWS:

ARTICLE I

1. The taking of polar bears shall be prohibited except as provided in Article III.
2. For the purpose of this Agreement, the term "taking" includes hunting, killing and capturing.

ARTICLE II

Each Contracting party shall take appropriate action to protect the ecosystems of which polar bears are a part, with special attention to habitat components such as denning and feeding sites and migration patterns, and shall manage polar bear populations in accordance with sound conservation practices based on the best available scientific data.

ARTICLE III

1. Subject to the provisions of Articles II and IV, any Contracting Party may allow the taking of polar bears when such taking is carried out:
   a) for bona fide scientific purposes; or
   b) by that Party for conservation purposes; or
   c) to prevent serious disturbance of the management of other living resources, subject to forfeiture to that Party of the skins and other items of value resulting from such taking; or
   d) by local people using traditional methods in the exercise of their traditional rights and in accordance with the laws of that Party; or
   e) wherever polar bears have or might have been subject to taking by traditional means by its nationals.
2. The skins and other items of value resulting from taking under subparagraphs (b) and (c) of paragraph 1 of this Article shall not be available for commercial purposes.

ARTICLE IV

The use of aircraft and large motorised vessels for the purpose of taking polar bears shall be prohibited, except where the application of such prohibition would be inconsistent with domestic laws.

ARTICLE V

A Contracting Party shall prohibit the exportation from, the importation and delivery into, and traffic within, its territory of polar bears or any part or product thereof taken in violation of this Agreement.

ARTICLE VI

1. Each Contracting Party shall enact and enforce such legislation and other measures as may be necessary for the purpose of giving effect to this Agreement.
2. Nothing in this Agreement shall prevent a Contracting Party from maintaining or amending existing legislation or other measures or establishing new measures on the taking of polar bears so as to provide more stringent controls than those required under the provisions of this Agreement.

ARTICLE VII

The Contracting Parties shall conduct national research programmes on polar bears, particularly research relating to the conservation and management of migrating polar bears populations, and exchange information on research and management programmes, research results and data on bears taken.

ARTICLE VIII

Each Contracting Party shall take actions as appropriate to promote compliance with the provisions of this Agreement by nationals of States not party to this Agreement.

ARTICLE IX

The Contracting Parties shall continue to consult with one another with the object of giving further protection to polar bears.

ARTICLE X

1. This Agreement shall be open for signature at Oslo by the governments of Canada, Denmark, Norway, the Union of Soviet Socialist Republics and the United States of America until 31 March 1974.
2. This Agreement shall be subject to ratification or approval by the signatory Governments. Instruments of ratification or approval shall be deposited with the Government of Norway as soon as possible.
3. This Agreement shall be open for accession by the Governments referred to in paragraph I of this Article. Instruments of accession shall be deposited with the Depository Government.
4. This Agreement shall enter into force ninety days after the deposit of the third instrument of ratification, approval of accession. Thereafter, it shall enter into force for a signatory or acceding Government on the date of deposit of its instrument of ratification, approval or accession.
5. This Agreement shall remain in force initially for a period of five years from its date of entry into force, and unless any Contracting party during that period request the termination of the Agreement at the end of that period, it shall continue in force thereafter.
6. On the request addressed to the Depository Government by any of the Governments referred to in paragraph I of this Article, consultations shall be conducted with a view to convening a meeting of representatives of the five Governments to consider the revision or amendment of this Agreement.
7. Any Party may denounce this Agreement by written notification to the Depository Government at any time after five years from the date of entry into force of this Agreement. The denunciation shall take effect twelve months after the Depository government has received this notification.
8. The Depository government shall notify the Governments referred to in paragraph I of this Article of the deposit of instruments of ratification, approval or accession, for the entry into force of this Agreement and of the receipt of notifications of denunciation and any other communications from a Contracting Party specially provided for in this Agreement.
9. The original of this Agreement shall be deposited with the Government of Norway which shall deliver certified copies thereof to each of the Governments referred to in paragraph I of this Article.
10. The Depository Government shall transmit certified copies of this Agreement to the Secretary-General of the United

Nations for registration and publication in accordance with Article 102 of the Charter of the United Nations.

The Agreement came into effect in May 1976, three months after the third nation required to ratify did so in February 1976. All five nations had ratified the Agreement by 1978. After the initial period of five years, all five Contracting Parties met in Oslo, Norway, in January 1981, and unanimously reaffirmed the continuation of the Agreement.

INDEX

Entries in *italics* refer to illustrations